High School Math — it's tricky

Math can be pretty tough — no question about that.

And as we sat around one afternoon, idly wondering what we could do to help the long-suffering youth out on the street, we thought hey, let's ease the furrowed brows and do some high school math.

So in putting this book together we've done everything we can to make things easier for you. We've gone through the standards with a fine-tooth comb, we've scrutinized Exam questions, we've found out everything you need to know, and we've stayed up late, night after night, drinking large cups of cocoa and working out the best way to explain things.

We've done everything we can to help you understand what's going on. We've detailed the methods with as many helpful hints as you could possibly want. And we've even tried to put in some fun stuff to keep you awake — in short, we've put our "all" into this book.

We've done our part — the rest is up to you.

What CGP's All About

Our sole aim here at CGP is to produce the highest quality books — carefully written, immaculately presented, and dangerously close to being funny.

Then we work our socks off to get them out to you — at the cheapest possible prices.

Contents

Section One — Numbers Mostly

Section Two — Shapes and Measurement

Section Three — Graphs

Section Four — Algebra

Section Five — Probability and Statistics

Published by Coordination Group Publications, Inc.

Contributors:
Simon Little
Andy Park
Glenn Rogers
Claire Thompson

With thanks to Marion Wright for proofreading.

ISBN 1-84146-852-5

Groovy website: www.cgpstudy.com

Printed by Johnson Printing, Boulder, CO.

Text, design, layout, and illustrations © Coordination Group Publications, Ltd. 2001, 2003
All rights reserved.

0403

Multiples and Factors

If you want to leave school, you have to pass that Exam. You gotta learn all this stuff.

Multiples

The MULTIPLES of a number are simply its TIMES TABLE:

e.g. the multiples of 13 are 13 26 39 52 65 78 91...

"Least Common Multiple" (LCM)

"Least Common Multiple" — sure, it sounds kind of complicated, but all it means is this:

The SMALLEST number that will DIVIDE BY ALL the numbers in question.

Method:
1) LIST the MULTIPLES of ALL the numbers.
2) Find the SMALLEST one that's in ALL the lists.
3) Piece of cake, isn't it?

Example: "Find the least common multiple (LCM) of 3 and 8."

Multiples of 3 are: 3, 6, 9, 12, 15, 18, 21, (24), 27 ...
Multiples of 8 are: 8, 16, (24), ...

So the least common multiple (LCM) of 3 and 8 is 24. Told you it was easy.

Factors

The FACTORS of a number are all the numbers that DIVIDE INTO IT. There's a special way to find them:

Example: "Find ALL the factors of 24."

1) Write the numbers 1, 2, 3, etc. in a column — like this:
2) Try dividing by each number in turn and write the answer in a second column. (Put a dash if it doesn't divide exactly.)
3) When you reach a number in the 1st column that you've already written in the 2nd column, stop.

Increasing by 1 each time

1 × 24
2 × 12
3 × 8
4 × 6
5 × –
6 × 4

So the FACTORS OF 24 are 1, 2, 3, 4, 6, 8, 12, 24

This method guarantees you find them ALL — but don't forget 1 and 24!

"Greatest Common Factor" (GCF)

The BIGGEST number that'll DIVIDE INTO ALL the numbers.

Method:
1) LIST the FACTORS of all the numbers.
2) Find the BIGGEST one that's in ALL the lists.
3) Cake.

Example: "Find the greatest common factor (GCF) of 54 and 72."

Factors of 54 are: (1), (2), (3), (6), (9), (18), 27, 54
Factors of 72 are: (1), (2), (3), 4, (6), 8, (9), 12, (18), 24, 36, 72

So the greatest common factor (GCF) of 54 and 72 is 18.

The Acid Test:

LEARN what Multiples and Factors are, AND HOW TO FIND THEM. Turn over and write it down.

1) List the first 10 multiples of 7 and the first 10 multiples of 9. Write down the LCM of 7 and 9.
2) List all the factors of 36 and all the factors of 84. Write down the GCF of 36 and 84.

Prime Numbers and Prime Factors

1) Basically, PRIME Numbers Don't Divide by Anything

And that's the best way to think of them.
So Prime Numbers are all the numbers that DON'T come up in Times Tables:

2 3 5 7 11 13 17 19 23 29 31 37 ...

As you can see, they're an awkward-looking bunch (that's because they don't divide by anything!). For example:

The only numbers that multiply to give 7 are	1 × 7
The only numbers that multiply to give 31 are	1 × 31

In fact the only way to get ANY PRIME NUMBER is 1 × ITSELF

OK, so a prime does divide by SOMETHING at least. But only 1 and itself. That's the lot.

2) They All End in 1, 3, 7, or 9 *(pretty much)*

1) 1 is NOT a prime number.

2) The first four prime numbers are 2, 3, 5, and 7.

3) 2 and 5 are the EXCEPTIONS because all the rest end in 1, 3, 7, or 9.

4) But NOT ALL numbers ending in 1, 3, 7, or 9 are primes, as shown here: (Only the circled ones are primes.)

(2)	(3)	(5)	(7)	9
(11)	(13)		(17)	(19)
21	(23)		27	(29)
(31)	33		(37)	39
(41)	(43)		(47)	49
51	(53)		57	(59)
(61)	63		(67)	69

Prime Factors are Factors that are Prime Numbers

Any number can be broken down into a string of PRIME NUMBERS all multiplied together — this is called "expressing it as a product of prime factors," and to be honest it's pretty tedious — but math teachers just love to give you questions on it, and it's not difficult so long as you know what it is.

The Factor Tree Method:

The mildly entertaining "Factor Tree" method is best, where you start at the top and split your number off into factors as shown.

Each time you get a prime, you circle it, and you finally end up with all the prime factors circled — which you can then arrange in order.

Example:

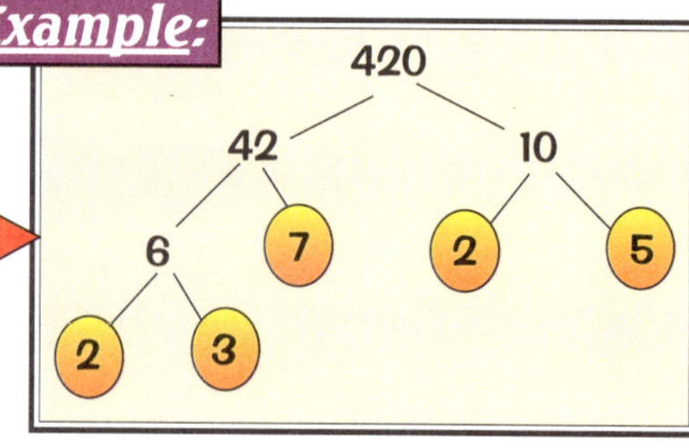

So, "as a product of prime factors," 420 = 2 × 2 × 3 × 5 × 7
(or $2^2 \times 3 \times 5 \times 7$).

The Acid Test: LEARN the main points in ALL 3 SECTIONS above.

Now cover the page and write down everything you've just learned.
1) Write down the first 15 prime numbers (*without* looking them up).
2) Express these numbers as products of prime factors: a) 990 b) 160 c) 1260

Fractions

Terrifyingly, you may be forced to demonstrate your prowess at fractions *in the Exam*. Better learn this then...

The Four Crucial Fraction Methods:

1) Multiplying — easy

Multiply top and bottom separately: $\frac{3}{5} \times \frac{4}{7} = \frac{3 \times 4}{5 \times 7} = \frac{12}{35}$

2) Dividing — quite easy

Turn the 2nd fraction *UPSIDE DOWN* and then *multiply*: $\frac{3}{4} \div \frac{1}{3} = \frac{3}{4} \times \frac{3}{1} = \frac{9}{4}$

3) Adding, subtracting — scary

Add or subtract *TOP LINE ONLY* — but *only if the bottom numbers are the same*:
(If they're not the same it gets very tricky — see the bit on common denominators below.)

$\frac{2}{6} + \frac{1}{6} = \frac{3}{6}$ $\quad \frac{5}{7} - \frac{3}{7} = \frac{2}{7}$

4) Canceling down — easy

Divide top and bottom by the same number,
'til they won't go any further:

$\frac{18}{24} \overset{\div 3}{=} \frac{6}{8} \overset{\div 2}{=} \frac{3}{4}$

Find the Least Common Denominator by Factoring

Don't attempt to add (or subtract) fractions unless the denominators (bottom nos.) are the same. You need to put both fractions over a *common denominator*. Any common denominator will do, but they'll probably ask you to find the *"least common denominator"* — so here's how you do it:

Example: Find $\frac{5}{6} + \frac{3}{20}$

Method:

1) *Rewrite* the denominators in their *prime factored form*: (See P.2)
$\frac{5}{6} + \frac{3}{20} = \frac{5}{2 \times 3} + \frac{3}{2 \times 2 \times 5}$

2) Write down the *first denominator*: 2×3

3) Write down any *extra factors* from the *second denominator* that you haven't already written down from the first. This is the *least common denominator*.
$2 \times 3 \times 2 \times 5$
You need to put in this extra "2" because there's a "2 x 2" in the second fraction.

4) Put *both fractions over the least common denominator* by multiplying the top and bottom of each fraction by the extra factors in the common denominator:
$\frac{5 \times 2 \times 5}{2 \times 3 \times 2 \times 5} + \frac{3 \times 3}{2 \times 2 \times 5 \times 3}$

5) Do the addition (see above), then cancel if you can.
(You can't in this example.)
$= \frac{50}{60} + \frac{9}{60} = \frac{59}{60}$

The Acid Test:

LEARN the Four Crucial Fraction Methods and how to get the same number on the bottom of fractions. Turn over and write it all down.

1) Cover the page and find: a) $\frac{5}{2} \times \frac{7}{4}$ b) $\frac{2}{11} \div \frac{3}{5}$ c) $\frac{2}{3} + \frac{1}{3}$ d) $\frac{5}{6} - \frac{1}{6}$ e) $\frac{3}{4} + \frac{9}{14}$ f) $\frac{4}{10} + \frac{4}{25}$

2) a) Simplify these fractions, then b) put them in order (smallest first): $\frac{12}{4}$, $\frac{6}{36}$, $\frac{56}{96}$, $\frac{13}{52}$, $\frac{17}{34}$

Percents

Percent problems are all the same... well, nearly. But they are similar enough that there are just *four distinct types* of questions you can be asked. Obviously then, it's pretty *essential* that you can:

1) Distinguish between the four types.
2) Remember the METHOD for each of them.

Type 1 — THESE ARE IDENTIFIED BY THE "%" SYMBOL IN THE QUESTION

This is the easiest type — they're always in the form:

FIND "something" % OF "something else"

For example: *"Find 15% of $25."*

Method:

1) *WRITE*: 15% OF $25

2) *TRANSLATE*: $\frac{15}{100} \times 25 = \$3.75$

3) *CHECK* THAT IT'S A *SENSIBLE ANSWER*.

Remember:
1) "OF" means "×"
2) "PERCENT" means "OUT OF 100," so 15% *means* "15 out of 100" — i.e. $\frac{15}{100}$

Example: A door-to-door salesman gets 28% commission on all sales. If he sells 20 dictionaries for $12 each, how much commission will he earn?

Answer: Total selling price = 20 × $12 = $240
Commission = 28% — so find 28% of $240: $\frac{28}{100} \times \$240 = \$67.20$

Type 2 — THESE ARE IDENTIFIED BY THEM NOT GIVING YOU THE "ORIGINAL VALUE"

These are the type most people get wrong — but only because they don't recognize them as a Type 2 and don't apply this simple method:

Example:
"A house increases in value by 20% to $180,000. Find its value before the increase."

Method:

$180,000 = 120%
÷ 120
$1500 = 1%
× 100
$150,000 = 100%

So the original price was $150,000

An *increase* of 20% means that $180,000 represents *120% of the original* value.

If it was a *DROP* of 20%, then you'd put *"$180,000 = 80%"* instead, and then divide by *80* rather than 120.

Always set them out exactly like this example. The trickiest part is deciding the top % figure on the right-hand side — the 2nd and 3rd rows are *always* 1% and 100%.

Concept — Number Sense

Convert fractions to decimals and percents and use these representations in estimations, computations, and applications.

Calculate the percentage of increases and decreases of a quantity.

Percents

Type 3 — THESE ARE IDENTIFIED BY THE WORD "PERCENT" IN THE QUESTION

These are always of the form:

EXPRESS "one thing" AS A PERCENT OF "another"

For example: *"Express 35¢ as a percent of $2.80."*

Method — FDP:

FDP: Fraction – Decimal – Percent

(See P.7 for more on this)

$\frac{35}{280} \xrightarrow{35 \div 280} 0.125 \xrightarrow{\times 100} 12.5\%$

Make a fraction using the two numbers — always with the smallest on top.

Divide them to get a decimal.

Then multiply by 100 to get a percent.

Percent of Change (an important example of Type 3)

You might have to give a change in value as a percent.
This is the formula you'll need — LEARN IT, AND USE IT:

$$\text{PERCENT OF "CHANGE"} = \frac{\text{"CHANGE"}}{\text{ORIGINAL}} \times 100$$

"Change" can mean all sorts of things such as: "profit," "loss," "appreciation," "depreciation," "increase," "decrease," "error," "discount," etc.

For example, percent of "profit" $= \frac{\text{"profit"}}{\text{original}} \times 100$

Don't forget the monster importance of using the ORIGINAL VALUE in this formula.

The Acid Test:

LEARN the first 3 Types, how you identify them, and the Method for each. Also LEARN the Formula for Percent of Change.

Now turn over and write down all the details you've just learned.

Identify the following questions as Type 1, 2, or 3, and apply the method for each.
Practice until you can do them without the notes:

1) A trader buys watches for $5 and sells them for $7.
Find his profit in $ and then express it as a percent.
2) Find the total cost of a plumber's bill given as: "$36 + 7.5% tax."
3) A car depreciates by 30% to $14,350. What was it worth before?
4) If a rubber llama's head costs $34.50 and sales tax is added at 8%, how much would you have to pay for it?

Percents — Compound Interest

Type 4 — Compound Interest

— IDENTIFIED BY THEM GIVING YOU AN ANNUAL RATE OF INTEREST AND ASKING YOU TO CALCULATE 3 YEARS' WORTH. Or something similar.

Method:

1) Find the interest after 1 year.
2) Add it on to the original amount.
3) Find the interest on the new total.
4) Add the interest on to the new total.
5) Etc.

Example: *"Paul has $1000 in a savings account that gives an annual rate of interest of 5%. If he leaves the account alone, how much money will he have in the account at the end of two years?"*

Answer:

1) Interest after 1 year: 5% of $1000 = $\frac{5}{100} \times \$1000 = \$50$
2) Add it on: $1000 + $50 = $1050
3) Interest after 2 years: 5% of $1050 = $\frac{5}{100} \times \$1050 = \$52.50$
4) Add it on: $1050 + $52.50 = $1102.50

SO THE ANSWER IS $1102.50 — EASY.

Harder Example — interest compounded twice a year:

"Sam has $20,000 in a savings account with an annual rate of interest of 1%, compounded twice a year. If she leaves the account alone, how much money will she have in her account at the end of 1 year?"

Answer: Eeek — interest compounded twice a year, so you have to halve the rate of interest, then work it out after 6 months, then after a year.

Step Zero: 6-monthly interest rate = 0.5%

1) Interest after 6 months: 0.5% of $20,000 = $\frac{1}{200}$ $\left(= \frac{0.5}{100}\right)$ $\times \$20,000 = \$100$
2) Add it on: $20,000 + $100 = $20,100
3) Interest after one year: 0.5% of $20,100 = $\frac{1}{200} \times \$20,100 = \$100.50$
4) Add it on: $20,100 + $100.50 = $20,200.50

SO THE ANSWER IS $20,200.50 — TRICKY, HUH.

The Acid Test: LEARN THE METHOD for finding COMPOUND INTEREST. Turn over and write it down.

1) Find the amount in Paul's account after 2 years if the annual rate of interest had been 10%.
2) Find the amount in Sam's account after a year if the annual rate of interest had been 5%.

Fractions, Decimals, and Percents

The one word that could describe all three of these is ***PROPORTION***. Fractions, decimals, and percents are simply *3 different ways* of expressing a *proportion* of something — and it's pretty important you should see them as *closely related and completely interchangeable* with each other.

This table shows the really common conversions that you should know off the top of your head:

Fraction	Decimal	Percent
$\frac{1}{2}$	0.5	50%
$\frac{1}{4}$	0.25	25%
$\frac{3}{4}$	0.75	75%
$\frac{1}{3}$	0.3333...	33%
$\frac{2}{3}$	0.6666...	67%
$\frac{1}{10}$	0.1	10%
$\frac{2}{10}$	0.2	20%
$\frac{x}{10}$	0.x	x0%
$\frac{1}{5}$	0.2	20%
$\frac{2}{5}$	0.4	40%

The more of those conversions you learn, the better — but for those that you *don't know*, you must *also learn* how to *convert* between the three types. These are the methods:

Fraction —Divide→ **Decimal** —× by 100→ **Percent**
e.g. $\frac{1}{2}$ is $1 \div 2$ = 0.5; e.g. 0.5×100 = 50%

Fraction ←The tricky one— **Decimal** ←÷ by 100— **Percent**
$0.75 = \frac{75}{100} = \frac{3}{4}$ ← = 0.75 ← = 75 ÷ 100 ← e.g. 75%

Converting Decimals to Fractions — a simple rule:

Converting decimals to fractions is only possible for *exact decimals* that haven't been rounded off. It's simple enough, but it's best shown by examples, so look at these and figure out the simple rule:

Examples:

$0.6 = \frac{6}{10}$ $0.3 = \frac{3}{10}$ $0.7 = \frac{7}{10}$ $0.x = \frac{x}{10}$ etc.

$0.12 = \frac{12}{100}$ $0.78 = \frac{78}{100}$ $0.45 = \frac{45}{100}$ $0.05 = \frac{5}{100}$ etc.

$0.345 = \frac{345}{1000}$ $0.908 = \frac{908}{1000}$ $0.024 = \frac{24}{1000}$ $0.xyz = \frac{xyz}{1000}$ etc.

These can then be *canceled down* (see P.3).

The Acid Test:

Cover the page and write out the top FDP table from memory, and then the rules for converting between fractions, decimals, & percents...

Then fill in all the spaces in this table:

Fraction	Decimal	Percent
$\frac{1}{5}$		
	0.35	
		45%
	0.12	
$\frac{1}{8}$		
	0.77	

Squares, Square Roots, and Reciprocals

Square Numbers

"Squared" means "multiplied by itself": $P^2 = P \times P$

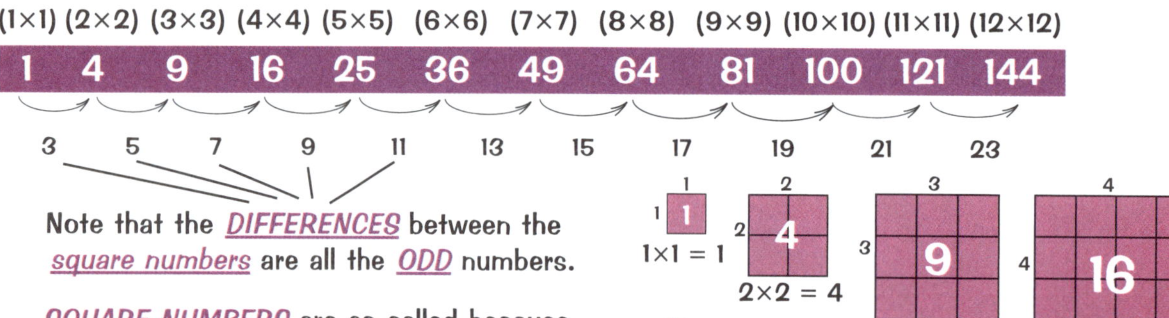

Note that the DIFFERENCES between the square numbers are all the ODD numbers.

SQUARE NUMBERS are so called because they're the areas of this pattern of squares:

1×1 = 1 2×2 = 4 3×3 = 9 4×4 = 16

The best thing you can do with square numbers is LEARN THEM — so you can spot them easily.

Square Roots

You know that "squared" means "multiplied by itself" — SQUARE ROOT is the opposite process.

The best way to think of it is this:

"Square Root" means "What Number Multiplied by Itself gives..."

Example 1 — a "perfect square"

A "perfect square" is where it's a square number, e.g. "Find the square root of 49." (i.e. "Find $\sqrt{49}$")

To do this you should say it as: "What number MULTIPLIED BY ITSELF gives... 49?"

(Of course you'll know almost instantly that the answer is 7.)

Example 2 — not a "perfect square"

"Find the square root of 70." (i.e. "Find $\sqrt{70}$")

These are hard without a calculator — but you can get a rough answer pretty easily.

$\sqrt{70}$ must be bigger than 8, since $8^2 = 64$, but less than 9, because $9^2 = 81$.

So $\sqrt{70}$ is somewhere between 8 and 9.

Reciprocals

"Taking a reciprocal" just means switching the top and bottom lines (or raising something to the power -1 — see P.9).

Examples: Find the reciprocals of a) 7, b) $\frac{2}{5}$.

a) The reciprocal of 7 is $7^{-1} = \frac{1}{7}$.   b) The reciprocal of $\frac{2}{5}$ is $\left(\frac{2}{5}\right)^{-1} = \frac{5}{2}$.

Taking a reciprocal of a number is the same as dividing 1 by the number, e.g. the reciprocal of $\frac{3}{4}$ is $1 \div \frac{3}{4} = \frac{4}{3}$.

The Acid Test:

LEARN THE FIRST 12 SQUARE NOS (so you can write them out without looking). Then LEARN what SQUARE ROOTS and RECIPROCALS are.

1) Write down the following square numbers: a) 4^2 b) 14^2 c) 30^2
2) Find the square root of: a) 49 b) 400 c) 10,000
3) Find two consecutive whole numbers that these square roots lie between: a) $\sqrt{10}$ b) $\sqrt{150}$
4) Find the reciprocals of these numbers: a) 12 b) 1 c) $\frac{4}{7}$

Concept — Number Sense Add, subtract, multiply, and divide rational numbers and take positive rational numbers to whole-number powers. Understand negative whole-number exponents. Multiply and divide expressions involving exponents with a common base. Multiply, divide, and simplify rational powers by using exponent rules.
Concept — Algebra and Functions Interpret positive whole-number powers as repeated multiplication. Simplify and evaluate expressions that include exponents.
Algebra 1 — Students understand and use such operations as taking the opposite, finding the reciprocal, and taking a root. They understand and use the rules of exponents.

Powers (or Exponents)

Powers are a very useful shorthand:

"Exponent" is just the fancy math name for a power (the little number at the top).

$2 \times 2 \times 2 \times 2 \times 2 \times 2 \times 2 = 2^7$ ("Two to the power 7")
$7 \times 7 = 7^2$ ("Seven squared")
$6 \times 6 \times 6 \times 6 \times 6 = 6^5$ ("Six to the power 5")
$4 \times 4 \times 4 = 4^3$ ("Four cubed")

That part is easy to remember. Unfortunately, there are EIGHT SPECIAL RULES for powers that are not quite so easy, but you do need to know them:

The Eight Rules

The "big" number has to be the same in both terms — so you couldn't do this with something like $3^4 \div 2^5$.

1) When MULTIPLYING, you ADD the powers

e.g. $3^4 \times 3^6 = 3^{6+4} = 3^{10}$ $\quad 8^3 \times 8^5 = 8^{3+5} = 8^8$

2) When DIVIDING, you SUBTRACT the powers

e.g. $5^4 \div 5^2 = 5^{4-2} = 5^2$ $\quad 12^8 \div 12^3 = 12^{8-3} = 12^5$

3) When RAISING one power to another, you MULTIPLY the powers

e.g. $(3^2)^4 = 3^{2\times4} = 3^8$ $\quad (5^4)^6 = 5^{24}$ $\quad 4^3 = (2^2)^3 = 2^6$ $\quad (x^3)^4 = x^{12}$

4) $X^1 = X$, ANYTHING TO THE POWER 1 is just ITSELF

e.g. $3^1 = 3$ $\quad 6 \times 6^3 = 6^4$ $\quad 4^3 \div 4^2 = 4^{3-2} = 4^1 = 4$

5) $X^0 = 1$, ANYTHING TO THE POWER 0 is just 1

e.g. $5^0 = 1$ $\quad 67^0 = 1$ $\quad 3^4 \div 3^4 = 3^{4-4} = 3^0 = 1$

6) $1^x = 1$, 1 TO ANY POWER is still just 1

e.g. $1^{23} = 1$ $\quad 1^{89} = 1$ $\quad 1^2 = 1$ $\quad 1^{1012} = 1$

7) FRACTIONS — Apply power to both TOP and BOTTOM

e.g. $\left(1\frac{3}{5}\right)^3 = \left(\frac{8}{5}\right)^3 = \frac{8^3}{5^3} = \frac{512}{125}$ $\quad \left(\frac{a}{b}\right)^3 = \frac{a^3}{b^3}$

8) NEGATIVE POWERS — Turn UPSIDE-DOWN and make the power POSITIVE

e.g. $7^{-2} = \frac{1}{7^2} = \frac{1}{49}$ $\quad \left(\frac{3}{5}\right)^{-2} = \left(\frac{5}{3}\right)^{+2} = \frac{5^2}{3^2} = \frac{25}{9}$ $\quad 4m^{-3} = \frac{4}{m^3}$

The "-3" only applies to the m, so the 4 stays on top.

The Acid Test:

LEARN the Eight Rules for Powers. Then turn over and write them all down. Keep trying until you can do it.

Now apply the rules to SIMPLIFY these:

1) a) $3^2 \times 3^6$ b) $4^3 \div 4^2$ c) $(8^3)^4$ d) $(3^2 \times 3^3 \times 1^6) \div 3^5$ e) $7^3 \times 7 \times 7^{-2}$

2) a) $5^2 \times 5^7 \times 5^3$ b) $1^3 \times 5^0 \times 6^2$ c) $5^{-2} \times 3^{-3}$ d) $\left(\frac{4}{7}\right)^2$

3) a) $(4^3 \times 2^2 \times 4^2) \div \left(2^3 \times \left(\frac{1}{2}\right)^{-3}\right)$ b) $x^{-1}(xy)^2$ c) $x^{-4}y^2 \div x^3y^{-2}$

Scientific Notation

What it Actually is:

A number written in Scientific Notation must *ALWAYS* be in *EXACTLY* this form:

This *number* must *always* be *BETWEEN 1 AND 10*.

This number is just the *NUMBER OF PLACES the Decimal Point (D.P.) moves*.

Learn the Three Rules:

1) The *front number* must always be *BETWEEN 1 AND 10.*

2) The power of 10, n, is simply: *HOW FAR THE D.P. MOVES.*

3) n is *positive for BIG numbers*, n is *negative for SMALL numbers.*

(This is much better than rules based on which way the D.P. moves.)

Examples:

1) *"Express 35,600 in scientific notation."*

METHOD:
1) Move the D.P. until 35,600 becomes 3.56 ($\times 10^n$).
2) The D.P. has moved 4 places so n = 4, giving: 10^4.
3) 35,600 is a BIG number so n is +4, not -4.

ANSWER: 3.5600. = 3.56×10^4

2) *"Express 8.14×10^{-3} as an ordinary number."*

METHOD:
1) 10^{-3} tells us that the D.P. must move 3 places...
2) ...and the "-" sign tells us to move the D.P. to make it a SMALL number (i.e. 0.00814, rather than 8140).

ANSWER: .008.14 = 0.00814

Compare by Looking at the 10^n Part

Example: *"Which of these numbers is bigger: 3.7×10^3 or 1.1×10^4?"*

METHOD:
1) You only need to look at the 10^n part — 10^4 is bigger than 10^3.
2) So 1.1×10^4 is bigger than 3.7×10^3.

It works every time — if you don't believe me, try a few examples. If you're still not convinced, learn it anyway.

The Acid Test:

LEARN the Three Rules, then turn over and write them down.
Now cover up the rest of this page and answer these:

1) What are the Three Rules for scientific notation?
2) Express 958,000 in scientific notation.
3) And the same for 0.00018.
4) Express 4.56×10^3 as an ordinary number.
5) Put these numbers in order, smallest first: 2.5×10^{-2}, 9.1×10^6, 5.0×10^{-3}, 1.1×10^7
6) Work this out: $3.2 \times 10^{12} \div 1.6 \times 10^{-9}$, and write down the answer, first in scientific notation and then as an ordinary number. (Tricky one, that.)

Concept — Number Sense

Add, subtract, multiply, and divide rational numbers (integers, fractions, and terminating decimals) and take positive rational numbers to whole-number powers.
Understand the meaning of the absolute value of a number; interpret the absolute value as the distance of the number from zero on a number line; and determine the absolute value of real numbers.

Negative Numbers and Multiplying Letters

Everyone knows RULE 1, but sometimes RULE 2 applies instead, so make sure you know *BOTH* rules... ...AND when to use them.

Rule 1

Only to be used when...

+	+	makes	+
+	–	makes	–
–	+	makes	–
–	–	makes	+

1) Multiplying or dividing

e.g. $-2 \times 3 = -6$, $-8 \div -2 = +4$, $-4x \times -2 = +8x$

2) Two signs appear next to each other

e.g. $5 - -4 = 5 + 4 = 9$, $4 + -6 - -7 = 4 - 6 + 7 = 5$

If you're asked to "take the opposite" of something, it just means change the sign.

Rule 2

THE NUMBER LINE

Use this when ADDING OR SUBTRACTING:

e.g. *"Simplify $4x - 8x - 3x + 6x$."*

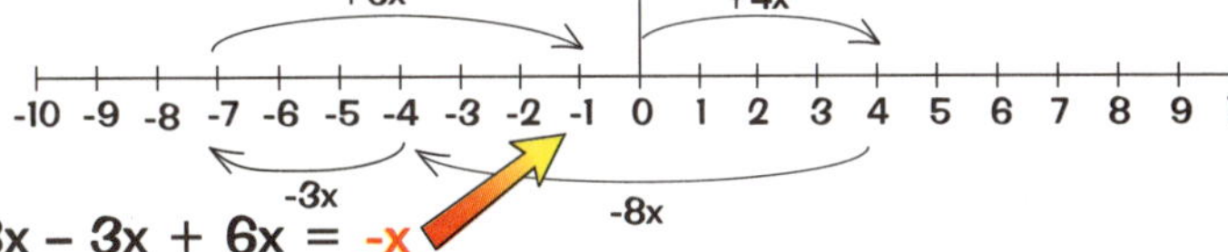

So $4x - 8x - 3x + 6x = -x$

|Absolute Value| — Ignore the Minus Sign

For more rib-tickling stuff on "absolute values" — *see P.54.*

① The absolute value of a number is just how big it is — regardless of whether it's positive or negative.

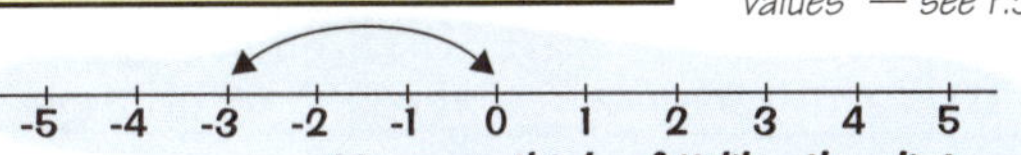

You can think of it like the distance from zero on the number line.

So the absolute value of -3 is 3. It's written like this: $|-3| = 3$

More examples (make sure you get them): $|-72\frac{1}{2}| = 72\frac{1}{2}$, $|7 - 15| = |-8| = 8$, $|+32| = 32$

② Watch out: Don't write $|x|$ or $|-x|$ as x. That's wrong: x might be negative, so this wouldn't be the absolute value. All you can do is leave them as $|x|$ *(unless you happen to know whether x is positive or negative)*.

Letters Multiplied Together

This is the superslick notation they like to use in algebra that just ends up making life difficult for you and me. You've got to remember these five rules:

1) " abc " means " $a \times b \times c$ " — The ×'s are often left out to make it clearer.

2) " gn^2 " means " $g \times n \times n$ " — Note that only the n is squared, not the g.

3) " $(gn)^2$ " means " $g \times g \times n \times n$ " — The parentheses mean that BOTH letters are squared.

4) " $p(q - r)^3$ " means " $p \times (q - r) \times (q - r) \times (q - r)$ " — Only the parentheses get cubed.

5) " $3n^2 \times 5np$ " can be written as " $15n^3p$ " — you just multiply all the bits together.

The Acid Test:

LEARN those rules for negative numbers and absolute values, and the 5 special cases of letters multiplied together.

1) Decide where Rule 1 and Rule 2 apply, and then find the value of the expression.
a) -4×-3 b) $-4 + -5 + 3$ c) $(3x + -2x - 4x) \div (2 + -5)$ d) $120 \div -40$

2) Simplify these absolute value expressions as far as possible:
a) $|-3|$ b) $|45|$ c) $|18 - 7|$ d) $|7 - 18|$ e) $|12x - 7x|$ f) $|7x - 12x|$

3) If m = 2 and n = -3, find: a) mn^2 b) $(mn)^3$ c) $m(4 + n)^2$ d) n^3 e) $3m^2n^3 + 2mn$

Review Summary for Section One

I know these questions seem difficult, *but they're the very best studying you can do*. The whole point of studying is to find out what you *don't* know, learn it, and then practice it until you do. These searching questions test how much you know *better than anything else ever can*. They follow the sequence of pages in Section One, so you can easily look up anything you don't know.

Keep learning these basic facts until you know them

1) What are the multiples of a number? What are the factors of a number?
2) Describe a method for finding the least common multiple of 10 and 25.
3) Are the factors of a number, x:
   a) all the numbers that divide into x, or b) all the numbers that divide by x?
4) What is a prime number?
5) What is a prime factor?
6) Which of these is a method that you could use to find the prime factors of a number:
   a) The "Factory" Method, b) The "Factor Tree" Method, or c) The "Lava Tree" Method?
7) Describe in words the 4 rules for doing fractions by hand (adding/subtracting, multiplying, etc.).
8) Describe a factoring method for finding the least common denominator of two fractions.
9) Describe the 4 types of percent questions and how to identify them.
10) Give details of the method for each of the 4 types of percent questions.
11) Give the formula for percent of change, and give 3 examples of it.
12) Describe a method for working out compound interest over 2 years, if the interest is compounded a) annually, and b) twice a year?
13) Explain the conversion methods between fractions, decimals, and percents.
14) List the first ten square numbers.
15) What does "square root" mean?
16) Describe how to find the reciprocal of a) a whole number, and b) a fraction.
17) List 8 rules of powers and give an example of each.
18) A number in scientific notation is $C \times 10^n$. Should the number C be between:
    a) a rock and a hard place, b) 1 and 10, or c) Beelzebub and the Atlantic Ocean?
19) Describe a method for figuring out which of these numbers is bigger: 1.2×10^7 and 130,000.
20) Two numbers are multiplied together. State whether the answer will be positive or negative if:
    a) both numbers are positive, b) both numbers are negative,
    c) one number is positive and the other is negative.
21) What does "absolute value" mean?
22) What's the superslick notation for writing multiplied letters, such as "$a \times z \times q$"?

Conversion Factors

You need *conversion factors* for problems involving things like *exchange rates* and *map scales*.

Method:

1) Find the Conversion Factor (always easy)
2) Multiply by it AND divide by it
3) Choose the common sense answer

Three Important Examples:

1) *"Convert 2.55 hours into minutes."* (This is NOT 2 hr 55 min)

1) Conversion Factor = 60 (simply because 1 hour = 60 min)
2) 2.55 hr × 60 = 153 min (makes sense)
   2.55 hr ÷ 60 = 0.0425 min (ridiculous answer!)
3) So clearly the answer is that 2.55 hr = 153 min. (= 2 hr 33 min)

2) *"If $1 = 9.65 Mexican pesos, how much is 47.36 pesos in $ and ¢?"*

1) Obviously, Conversion Factor = 9.65 (the "exchange rate")
2) 47.36 × 9.65 = $457.02
   47.36 ÷ 9.65 = $4.91
3) Not quite so obvious this time, but if roughly 10 pesos = $1, then 47 pesos can't be much — surely not $457, so the answer must be $4 and 91¢.

3) *"A map has a scale of 1:20,000. How big in real life is a distance of 3 inches on the map?"*

1) Conversion Factor = 20,000
2) 3 in × 20,000 = 60,000 in (looks OK)
   3 in ÷ 20,000 = 0.00015 in (not good)
3) So 60,000 inches is the answer.

How do you convert to yards?

To convert 60,000 inches to yards:

1) C.F. = 36 (inches ⟷ yards)
2) 60,000 × 36 = 2,160,000 yd (hmm...)
   60,000 ÷ 36 = 1667 yd (more like it)
3) So answer = 1667 yd.

The Acid Test:

LEARN the 3 steps of the Conversion Factor method. Then turn over and write them down.

1) Convert 23 minutes into hours.
2) Which is more, US$34 or CAN$46? (Use the conversion factor of US$1 = CAN$1.55)
3) A map is drawn to a scale of 1:250,000.
   A road is 8 km long. How many cm will this be on the map? (1 km = 100,000 cm)

Compare weights, capacities, geometric measures, times, and temperatures within and between measurement systems (e.g., miles per hour and feet per second, cubic inches to cubic centimeters).

Customary and Metric Units

This topic is *Easy Points* — make sure you get them.

Customary Units

1) Length — inches (in), feet (ft), yards (yd), miles (mi)
2) Area — square inches (in^2), square feet (ft^2), square yards (yd^2), square miles (mi^2)
3) Volume — cubic inches (in^3), cubic feet (ft^3), gallons (gal), pints (pt), quarts (qt)
4) Weight — ounces (oz), pounds (lb), tons (tn)
5) Speed — miles per hour (mph)

MEMORIZE THESE KEY FACTS:

1 foot = 12 inches
1 yard = 3 feet
1 gallon = 8 pints = 4 quarts
1 pound = 16 ounces (oz)
1 ton = 2000 pounds (lb)
1 mile = 1760 yards

Metric Units

1) Length — millimeters (mm), centimeters (cm), meters (m), kilometers (km)
2) Area — mm^2, cm^2, m^2, km^2
3) Volume — mm^3, cm^3, m^3, liters (l), milliliters (ml)
4) Weight — grams (g), kilograms (kg), metric tons
5) Speed — kilometers per hour (km/h), meters per second (m/s)

LEARN THESE TOO:

1 cm = 10 mm	1 (metric) ton = 1000 kg
1 m = 100 cm	1 liter = 1000 ml
1 km = 1000 m	1 liter = 1000 cm^3
1 kg = 1000 g	1 cm^3 = 1 ml

Customary-Metric Conversions

You should be given these in the Exam, but it's still a very good idea to LEARN THEM ANYWAY.

APPROXIMATE CONVERSIONS

1 kg ≈ 2.2 lb	1 gallon ≈ 3.8 liters
1 m ≈ 1.1 yards (or 1 yard + 10%)	1 foot ≈ 30 cm
1 liter ≈ 2.1 pints	1 ton ≈ 1 metric ton
1 inch ≈ 2.5 cm	1 mile ≈ 1.6 km or 5 miles ≈ 8 km

Using Customary-Metric Conversion Factors

Convert 45 mm into cm.

ANS: Conversion factor = 10, so × or ÷ by 10, which gives 450 cm or 4.5 cm. (sensible)

Fahrenheit and Celsius — Add or Subtract 32 as well

To convert temperatures from degrees Fahrenheit (°F) to degrees Celsius (°C), use these formulas:

$$F = (1.8 \times C) + 32$$

Multiply then add.

$$C = \frac{F - 32}{1.8}$$

Subtract then divide.

The Acid Test:

LEARN all the Conversion Factors in the boxes above then cover up the page and write them all down.

1) How many liters is 3.5 gallons?
2) Roughly how many meters is 200 yards?
3) A rod is 46 inches long. What is this in cm?
4) Convert 10°C to Fahrenheit.

Other Conversions and Scale Drawings

Do things like m/s in Two Stages

If there are 2 "parts" to the units — DO THE CONVERSION IN 2 STAGES.

Example: A car's traveling at 50 miles per hour. What is this speed in feet per second?

a) *Convert 50 miles per hour into feet per hour.*
There are 3 feet in a yard and 1760 yards in a mile, so conversion factor = 3 × 1760 = 5280.
So × or ÷ 50 mph by 5280, which gives:
0.0095 feet per hour (ridiculous) or 264,000 feet per hour. (sensible)

b) *Then convert 264,000 feet per hour into feet per second.*
60 seconds in a minute and 60 minutes in an hour, so conversion factor = 60 × 60 = 3600.
So × or ÷ 264,000 by 3600, which gives:
950,400,000 feet per second (not likely) or 73.3 feet per second. (sensible)

Scale Drawings and Models — Easy, but watch out for Areas

1) Scales WITHOUT UNITS (e.g. 1:5000) — use the number as a CONVERSION FACTOR.
2) If units are given, write the scale with the units you're converting FROM on the left.
3) DIVIDE FOR 1, AND MULTIPLY FOR ALL. (See P.18 for more info.)

Example: The scale of this drawing is 1 inch to 3 feet.

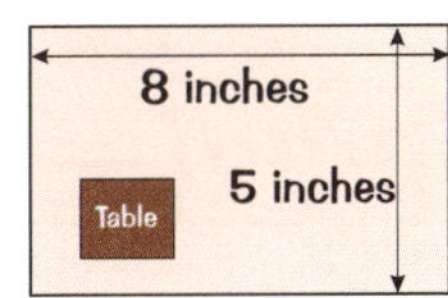

a) *How wide is the actual room?* Converting from inches to feet:
1 inch = 3 feet, so 8 inches = 8 × 3 feet = 24 feet wide.

b) *How long would a 7 foot long table be in the drawing?* Converting from feet to inches:
3 feet = 1 inch. So 1 foot = $\frac{1}{3}$ inches, and 7 feet = 7 × $\frac{1}{3}$ = 2.33 inches.

Divide to get "1 foot = ..."
...and then multiply to get "7 feet = ..."

c) *What's the area of the room?* A little more thought's required with this one...
Think of a square 1 inch by 1 inch — it represents an area of 3 feet × 3 feet = 9 square feet.

So for areas, the scale is: 1 in² to 9 ft² ← This is the square of the "length" scale.

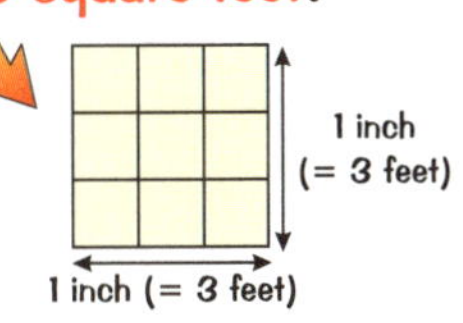

The area of the drawing is 8 in × 5 in = 40 square inches.
So the area of the room is 9 × 40 = 360 square feet.

Change of units — it's the same thing really...

The same thing applies to the conversion factor when you start to use different units, e.g. if you want to use centimeters instead of feet.

Length: 1 foot = 30 cm **BUT** Area: 1 ft² = 900 cm² (= 30²) Volume: 1 ft³ = 27,000 cm³ (= 30³)

The Acid Test:

PRACTICE those tricky double-unit conversions and scale drawing examples until you DON'T need to look back at the page.

1) A satellite is traveling at 24,000 miles per hour. How fast is this in kilometers per second?
2) A map is drawn using a scale of 1 mile = 2.5 inches. What distance on the map would represent 6 miles? How far from each other are two towns drawn 3.5 inches apart?
3) The density of water is 1000 kg/m³. What is this density in lb/yd³? (Careful...)

Density and Speed

You might think this is science, but they also teach it in Math (as if learning it once wasn't enough...). The standard formula for density is:

Density = Mass ÷ Volume

It's useful to put it in a FORMULA TRIANGLE like this:

How do you use a Formula Triangle?

1) COVER UP the thing you want to find and just WRITE DOWN what is left showing.
2) Now PUT IN THE VALUES for the other two things and WORK IT OUT.

One way or another you MUST remember this formula for density, because chances are they won't give it to you. The best method by far is to remember the order of the letters in the FORMULA TRIANGLE as MDV (think of "MeDieVal" — e.g. a medieval knight wearing dense armor).

EXAMPLE: *"Find the volume of an object which has a mass of 40 g and a density of 6.4 g/cm³."*

ANSWER: To find volume, cover up V in the formula triangle.
This leaves $\frac{M}{D}$ showing, so... V $= M \div D = 40 \div 6.4$
$= 6.25\text{ cm}^3$

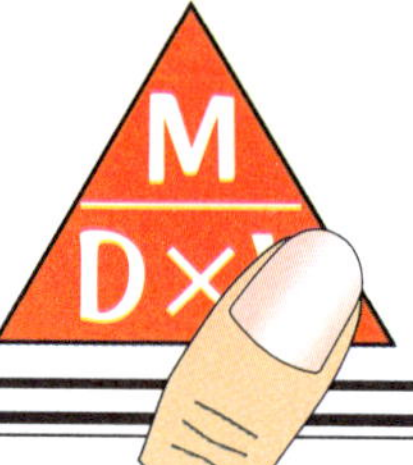

Speed = Distance ÷ Time

This is very common. In fact, people are always being given this problem — *and they generally don't give you the formula!* Either *learn it beforehand* or wave goodbye to *bunches of easy points*. Life isn't all bad though — there's an easy FORMULA TRIANGLE:

Of course, you still have to remember the order of the letters in the triangle (DST) — this time you could think of the word "DuST" to help you (e.g. a dust cloud left by something speeding past).

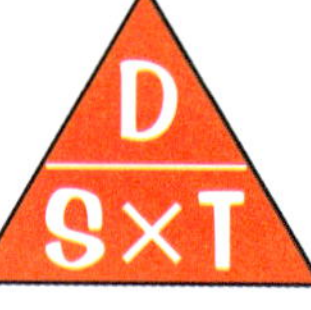

EXAMPLE: *"A car travels 90 miles at 36 miles per hour. How long does it take?"*

ANSWER: We want to find the TIME, so cover up T in the triangle which leaves D ÷ S,
so T = D ÷ S = distance ÷ speed = 90 ÷ 36 = 2.5 hours

LEARN THE FORMULA TRIANGLE, AND YOU'LL FIND QUESTIONS ON *SPEED, DISTANCE, AND TIME* VERY EASY.

The Acid Test:

LEARN the formulas for DENSITY and SPEED — and also the two Formula Triangles.

1) What's the formula triangle for density?
2) A metal object has a volume of 45 cm³ and a mass of 742.5 g. What is its density?
3) Another piece of the same metal has a volume of 36.5 cm³. What is its mass?
4) What's the formula for speed in terms of distance and time?
5) Find the time taken for a person walking at 3.2 mph to cover 24 miles.
Also, find how far she'll walk in 3 hrs 30 min.

Density and Speed

Break complicated problems into Easy Steps

Try using a *formula triangle* for *each separate part* of a more-involved problem.

EXAMPLE: *"Jane travels for 4 hours at an average speed of 30 miles per hour, then 3 hours at an average speed of 55 miles per hour. What is her average speed for the entire trip?"*

ANSWER: OK, it's the usual formula, but we need the *TOTAL distance* and the *TOTAL time*:

The total time's easy — it's 7 hours.

For the total distance, use D = S × T for *each part* of the journey:

1st part: D = 30 mph × 4 hr = 120 miles
2nd part: D = 55 mph × 3 hr = 165 miles

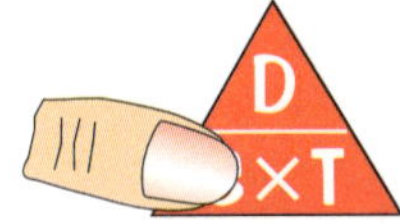

So total D = 120 + 165 = 285 miles

So Average Speed for Entire Trip = Total D ÷ Total T = 285 miles ÷ 7 hr
= 40.7 mph

It's a similar thing with average *density* — you just use the TOTAL mass and the TOTAL volume.

Units — getting them Right

By *units* we mean things like *mph, ft², in³, cm, m, m/s, km²*, etc., and as a rule you don't have to worry too much about them. However, when you're using a FORMULA TRIANGLE, there's one special thing you need to know. It's simple enough, *but you must know it:*

The UNITS you get OUT of a Formula DEPEND ENTIRELY upon the UNITS you put INTO IT

So for example if you put a *distance in cm* and a *time in seconds* into the formula triangle to work out speed, the answer must come out in *cm per second* (cm/s).
It's pretty simple when you think about it — as long as you take care with this kind of question:

Example *"A boy walks 800 m in 10 minutes. Find his speed in km/h."*

ANSWER: If you just do *"800 m ÷ 10 minutes"* your answer will be a speed, sure, but in *meters per minute* (m/min), which is no good at all.

Instead you must *CONVERT INTO KM AND HOURS* first:

800 m = *0.8 km*   10 min = *1/6* (mins ÷ 60).

Then you can divide *0.8 km* by *1/6 hours* to get *4.8 km/h*, which is much more like it.

The Acid Test:

LEARN how to yodel while tying your shoes, then sing the Star-Spangled Banner off-key.

1) Claire travels 3 miles at 1 mph, then discards her pogo stick and walks 7 miles at 3 mph. Find her average speed for the entire trip.
2) My pet turtle's shell has a volume of 250 cm³ and weighs 2 kg. His body (excluding shell) weighs 4 kg and has a volume of 1000 cm³. Find the average density of Bill (that's his name).
3) A sprinter can run 100 meters in 10 seconds. Find his average speed in kilometers per hour.

Divide for One, Then Multiply for All

There are lots of questions that at first sight seem completely different but in fact can all be done using one of two methods: The Golden Rule or The Silver Rule (P.19).

The best way to work out which you use is to get your head around the examples on these pages.

The Golden Rule says:

DIVIDE FOR ONE, THEN MULTIPLY FOR ALL

Example 1:

"5 quarts of sprout juice cost $13. How much will 3 quarts cost?"

ANSWER: *The Golden Rule* says: **DIVIDE FOR ONE, THEN MULTIPLY FOR ALL**

which means:

Divide the price by 5 to find how much FOR ONE QUART, then multiply by 3 to find how much FOR 3 QUARTS.

So... $13 ÷ 5 = 2.6 = $2.60 (for 1 quart)
× 3 = $7.80 (for 3 quarts)

Example 2:

"It takes a printer 20 minutes to print 400 stickers. Roughly how long will it take for the same machine to do 700 stickers?"

ANSWER: *The Golden Rule* says: **DIVIDE FOR ONE, THEN MULTIPLY FOR ALL**

which means:

Divide the time by 400 to find how long FOR ONE STICKER, then multiply by 700 to find how long FOR 700 STICKERS.

So... 20 ÷ 400 = 0.05 (minutes for 1 sticker)
× 700 = 35 (seconds for 700 stickers)

Example 3:

"45 radios were chosen at random from a batch of 3000. 3 of the radios were found to be damaged. Roughly how many of the 3000 radios would you expect to be damaged?"

ANSWER: *The Golden Rule* says: **DIVIDE FOR ONE, THEN MULTIPLY FOR ALL**

Divide the number of damaged radios by 45 to find how likely ONE RADIO is to be damaged, then multiply by 3000 to find how many damaged radios you're likely to find out of 3000 RADIOS.

So... 3 ÷ 45 = 0.0667 (damaged radios out of every 1 radio)
× 3000 = 200 (damaged radios out of 3000 radios)

(I know it doesn't make sense yet, but bear with me...)

Multiply for All, Then Divide

The Silver Rule is the Opposite of The Golden Rule

1) You use *The Golden Rule* when *one quantity increases as the other increases* (e.g. the more you buy, the more you pay).
2) You use *The Silver Rule* when *one quantity decreases as the other increases* (e.g. the more workers, the less time it takes).

The Silver Rule says:

MULTIPLY FOR ALL, THEN DIVIDE

Example 4: "It usually takes 8 people 5 days to bring in the harvest. How many people would be needed to bring in the harvest in 2 days?"

ANSWER: The Silver Rule says: **MULTIPLY FOR ALL, THEN DIVIDE**

Multiply the people by the days to find the number of PERSON-DAYS, then divide by 2 to find the number of people for it to take 2 days.

So... 8 people × 5 days = 40 person-days (So it would take 1 person 40 days, or 40 people 1 day.)
÷ 2 = 20 people (in 2 days)

Example 5: "It takes 8 toasters 3 hours to make enough toast for Andy's breakfast. How long would it take if he bought 10 more toasters?"

ANSWER: The Silver Rule says: **MULTIPLY FOR ALL, THEN DIVIDE**

Multiply the toasters by the hours to find the number of TOASTER-HOURS, then divide by 18 to find the number of hours for 18 toasters.

So... 8 toasters × 3 hours = 24 toaster-hours  (So it would take 1 toaster 24 hours, or 24 toasters 1 hour.)
÷ 18 = $1\frac{1}{3}$ hours (with 18 toasters)

The Acid Test:

Learn the *GOLDEN RULE* and the *SILVER RULE*, then learn *AT LEAST ONE EXAMPLE* of each.

1) A toy car takes 15 minutes to travel 27 yards. If it carries on at the same speed, how long will it take to reach 80 yards?
2) In a recent survey of 30 women in a health club, 11 said they preferred unicycling to wrestling. Another survey is going to be done with 500 women at a bigger health club. Predict how many of the 500 women will say they prefer unicycling to wrestling.
3) It usually takes about 12 hours for 700 bees to produce enough honey to fill a jar. Roughly how many bees would be needed to fill the jar in 7 hours?
4) There are 150 people at a birthday party. The cake is shared equally between all the people at the party, and each person gets 6 pieces of cake. How many pieces of cake would each person get if there had been just 80 people at the party?

Perimeters and Areas

OK, the bad news is that you won't always be given these formulas in the Exam.
But the good news is: that means you MUST LEARN ALL OF THESE and learning formulas is fun! (sorry)

Shape	Perimeter	Area
1) SQUARE 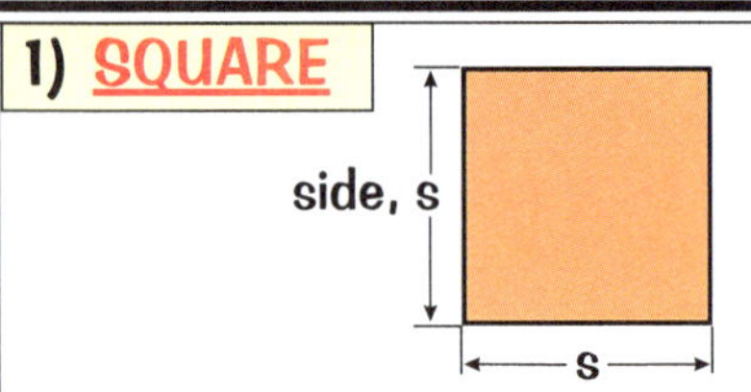	All sides the same length so PERIMETER = sum of all sides $= s + s + s + s$ $P = 4s$	AREA = length of side × itself $= s \times s$ $A = s^2$
2) RECTANGLE  	2 pairs of sides of the same length so PERIMETER = sum of all sides $= \ell + \ell + w + w$ $P = 2\ell + 2w$	AREA = length × width $= \ell \times w$ $A = \ell w$
3) PARALLELOGRAM 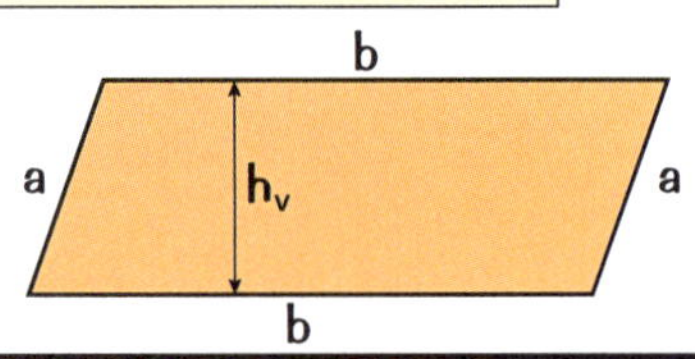	As above... PERIMETER = sum of all sides $= a + a + b + b$ $P = 2a + 2b$	AREA = base × vertical height $= b \times h_v$ $A = bh_v$
4) ANY TRIANGLE 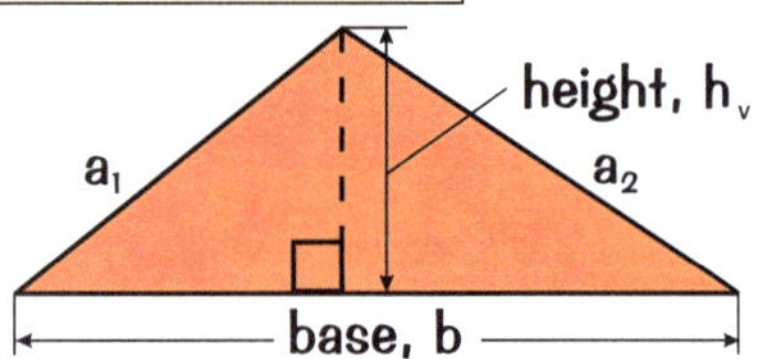	No 2 sides necessarily the same length so PERIMETER = sum of all sides $P = a_1 + a_2 + b$ For an *equilateral* triangle, $P = 3s$	AREA = ½ × base × height $= \frac{1}{2} \times b \times h_v$ $= \frac{1}{2}bh_v$ 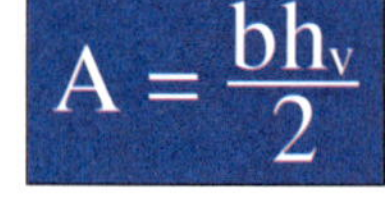 $A = \frac{bh_v}{2}$ Remember that the *height* must always be the *vertical height*, NOT the sloping height.
5) TRAPEZOID 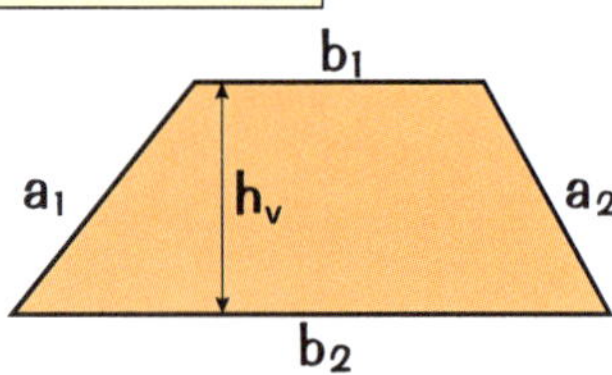	No 2 sides necessarily the same length so PERIMETER = sum of all sides $P = a_1 + a_2 + b_1 + b_2$	AREA = average length of parallel sides × perpendicular distance between them $= \{(b_1 + b_2) \div 2\} \times h_v$ $A = \frac{1}{2}(b_1 + b_2)h_v$
6) REGULAR POLYGONS 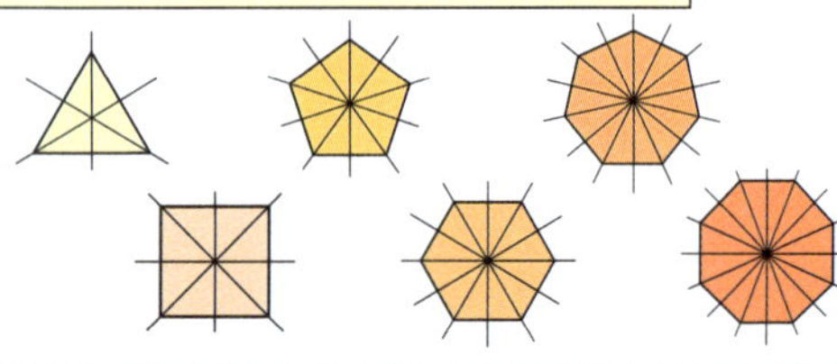	All sides the same length so PERIMETER = sum of all sides = side length (s) × No. of sides (n) $P = ns$	AREA = depends on shape You don't need to remember area formulas for anything this complicated.

The Acid Test: *Learn EVERYTHING on this page.*

Then turn over and write down all the details that you can remember. Then try again.

Perimeters and Areas

To deal with circle questions, you need to LEARN THE 2 FORMULAS — CIRCUMFERENCE and AREA. Once you've done that, the questions become so friendly and loveable, you'll want to cuddle them.

First of all, conquer an old fear...

1) π "A Number a Little Bigger than 3"

The big thing to remember is that π (called "pi") only seems confusing because it's a scary-looking Greek letter. In the end, it's just an ordinary number (3.14159...) that is rounded off to either 3 or 3.14 or 3.142 (depending on how accurate you want to be).

And that's all it is: *A NUMBER A LITTLE BIGGER THAN 3*.

2) Diameter is TWICE the Radius

The *RADIUS* goes from the center of the circle to the outside. The *DIAMETER* goes right across the circle, through the center, so it's *TWICE* the length of the *RADIUS*.

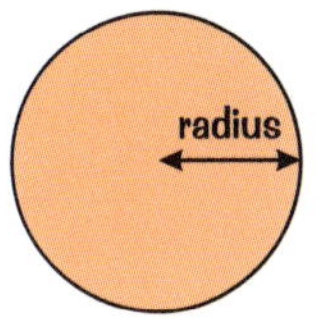

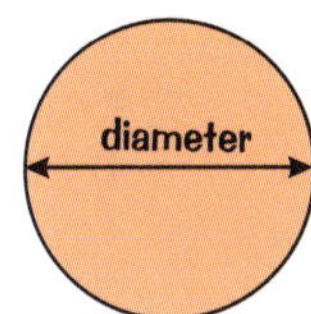

EXAMPLES:

If the radius is 4 in, the diameter is 8 in.
If the radius is 12 ft, the diameter is 24 ft.
If D = 12 cm, then r = 6 cm.
If diameter = 2 mm, then radius = 1 mm.

3) The Big Decision:

"*Which circle formula do I use?*"

WORKING OUT AREA OR CIRCUMFERENCE — there is a difference you know!

1) If you need to find the area of a circle, *YOU MUST* use this FORMULA FOR THE AREA:

$$A = \pi r^2$$

(i.e. $\pi \times r^2$)

(It makes no difference whether you know the radius or the diameter, because it's really easy to work out one from the other.)

2) If you need to find the distance around the circle (the circumference), *YOU MUST* use the FORMULA FOR CIRCUMFERENCE:

$$C = \pi D \quad (\text{or } 2\pi r)$$

EXAMPLE: *"Find the circumference and the area of the circle shown below."*

ANSWER: Radius = 5 in, so Diameter = 10 in (easy)

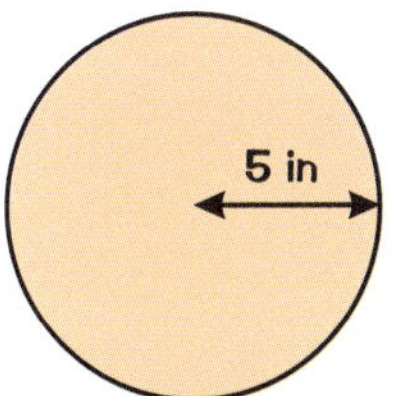

Formula for CIRCUMFERENCE is:
$C = \pi \times D$, so
$C = 3.14 \times 10$
$= 31.4$ in

Formula for AREA is:
$A = \pi \times r^2$
$= 3.14 \times (5 \times 5)$
$= 3.14 \times 25 = 78.5 \text{ in}^2$

The Acid Test:

There are 3 SECTIONS on this page. They're all mighty important — LEARN THEM.

Now cover the page and write down everything you've learned. Frightening isn't it.

1) A plate has a diameter of 14 cm. Find its area and circumference using the methods you've just learned. Remember to show all your work.
2) A circular flower bed has a radius of 6 yd. Find its area and circumference.

Perimeters and Areas

1) Perimeters of Complicated Shapes

Make sure you know these details about perimeter:

1) Perimeter is the distance all the way around the outside of a 2-D shape.

2) To find a PERIMETER, you ADD THE LENGTHS OF ALL THE SIDES, but... THE ONLY RELIABLE WAY to make sure you get all the sides is this:

1) Put a big blob at *ONE CORNER* and then go around the shape.
2) Write down the length of *EVERY SIDE* as you go.
3) Even sides that seem to have no length given — you *MUST* figure them out.
4) Keep going until you get back to the *BIG BLOB*.
5) *ADD UP* all the lengths.

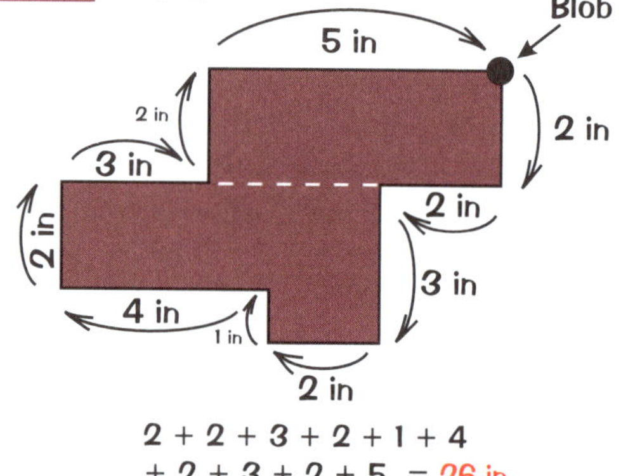

2 + 2 + 3 + 2 + 1 + 4
+ 2 + 3 + 2 + 5 = 26 in

Yes, I know you think it's yet another fussy method, but believe me, it's so easy to miss a side. You must use GOOD RELIABLE METHODS for EVERYTHING — or you'll miss things willy-nilly.

2) Areas of Complicated Shapes

1) SPLIT THEM UP into the 3 basic shapes: RECTANGLE, TRIANGLE, AND CIRCLE.
2) Work out the area of each part SEPARATELY.
3) Then ADD THEM ALL TOGETHER (or sometimes SUBTRACT them).

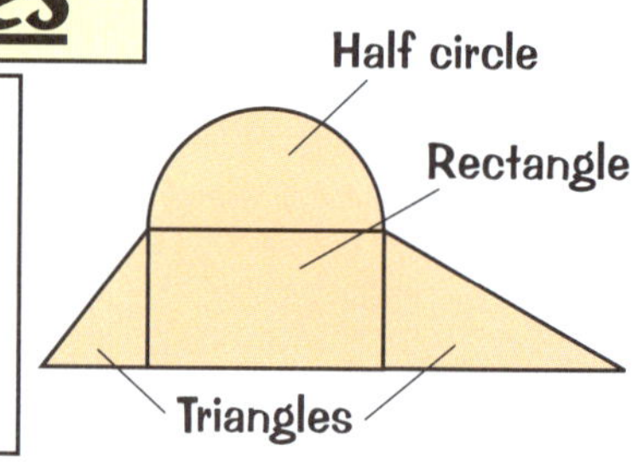

EXAMPLE: *"Work out the area of this shape:"*

ANSWER:

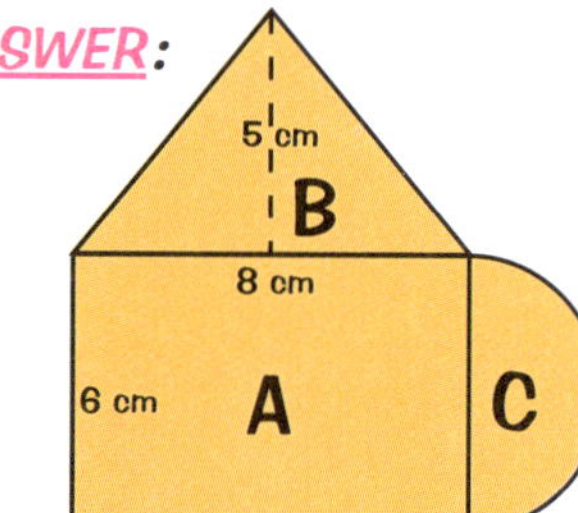

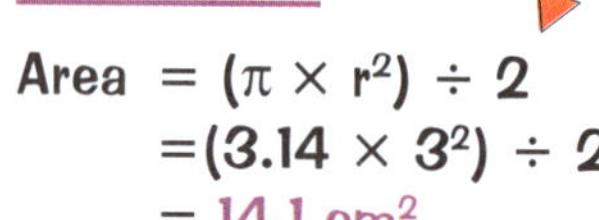

Rectangle A:

Area $= \ell \times w$
$= 8 \times 6$
$= 48\ \text{cm}^2$

Triangle B:

Area $= \frac{1}{2} \times b \times h_v$
$= \frac{1}{2} \times 8 \times 5$
$= 20\ \text{cm}^2$

Half Circle C:

Area $= (\pi \times r^2) \div 2$
$= (3.14 \times 3^2) \div 2$
$= 14.1\ \text{cm}^2$

TOTAL AREA = 48 + 20 + 14.1 = 82.1 cm²

The Acid Test:

LEARN THE RULES for finding the perimeter and area of complicated shapes.

1) *Turn over and write down* what you've learned.
2) Find the perimeter and area of the shape shown here:

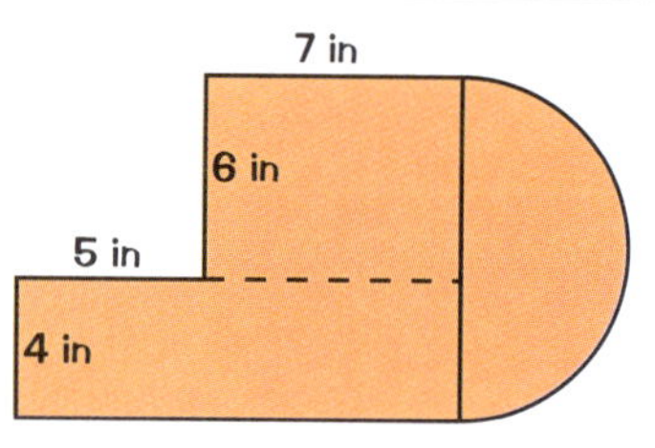

Volume

Volumes — You must Learn these Too!

Prisms

Now, for some reason, not many people know what a prism is, but they're important — so make sure YOU know.

A PRISM is a solid (3-D) object that has a constant area of cross-section — i.e. it's the same shape all the way through.

Triangular Prism

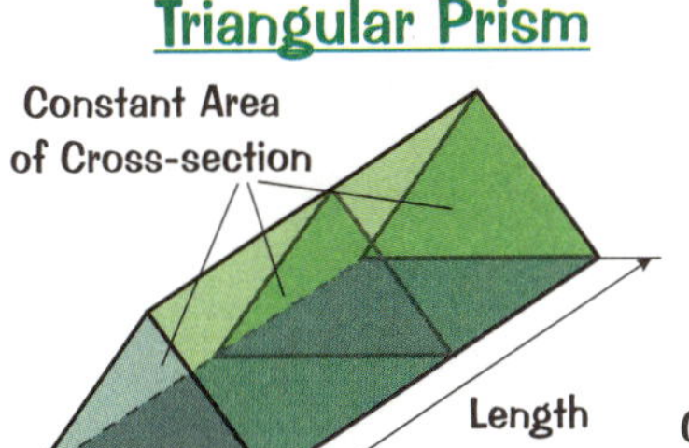

Hexagonal Prism

(a flat one, certainly, but still a prism)

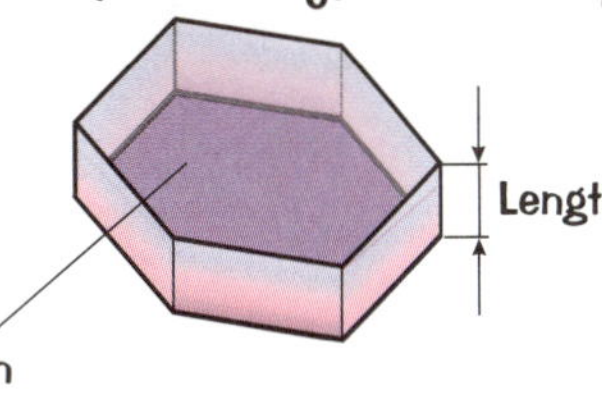

Circular Prism

(or Cylinder)

Constant Area of Cross-section

Length

$$\text{Volume of prism} = \text{Cross-sectional Area} \times \text{length}$$

As you can see, the formula for the volume of a prism is *very simple*. The *difficult* part, usually, is *finding the area of the cross-section*.

Rectangular Prism

This is the easy one — you know how to find the volume of a box — its cross-sectional area is just its width × its height.

Volume of Rectangular prism = length × width × height

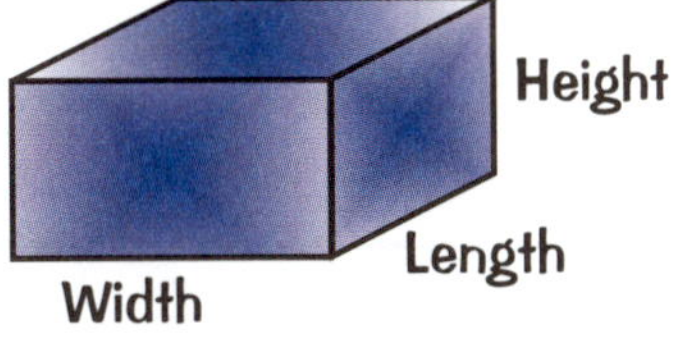

(The other word for volume is *CAPACITY*)

The Acid Test:

LEARN this page. Then turn over and try to write it all down. Keep trying until you can do it.

Find the volume of these two shapes:
What is the name of shape B?

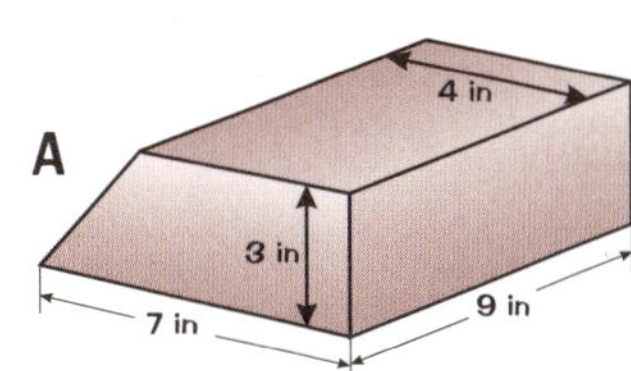

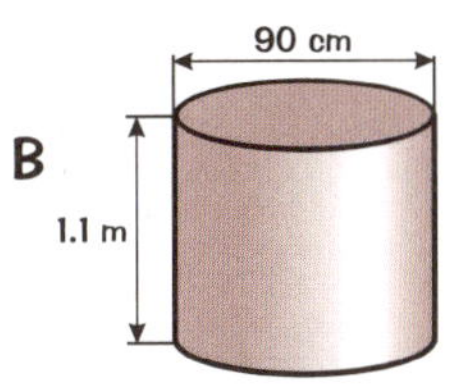

Surface Area

Like volume, SURFACE AREA only applies to solid 3-D objects, and it's simply *the total area of all the outer surfaces added together*. If you were painting the object, it's what you'd paint.

Surface Area and Nets

With most shapes, the best way to find the surface area is simply to *work out each side in turn and then* ADD THEM ALL TOGETHER.

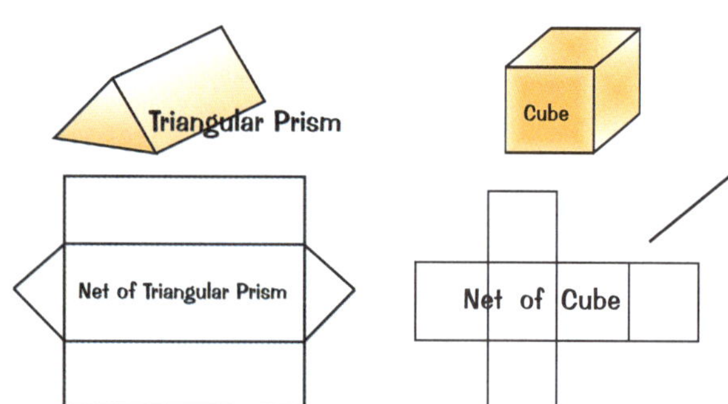

Thinking about the NET of the shape can help.

A net is just A SOLID SHAPE FOLDED OUT FLAT.

So obviously:
SURFACE AREA OF SOLID = AREA OF NET.

1) A triangular prism has 5 sides — 3 rectangles and 2 triangles.
   To find the surface area, just work out the area of each side separately and add them up.

2) All the sides on a cube are the same size and there are 6 sides, so you get this nice formula:

Surface area of a cube: $S = 6s^2$

Surface Area of a Cylinder

The surface area of a cylinder is a particularly important one.
(And the net is a particularly weird one...)

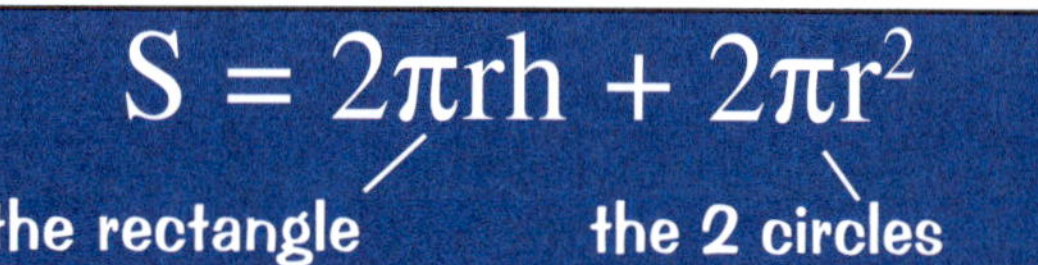

$$S = 2\pi rh + 2\pi r^2$$

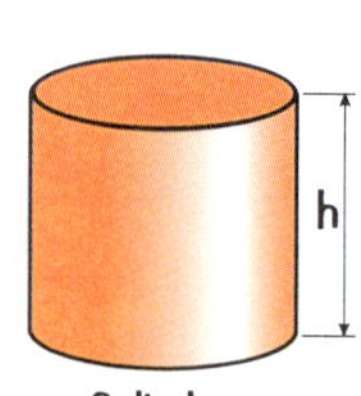

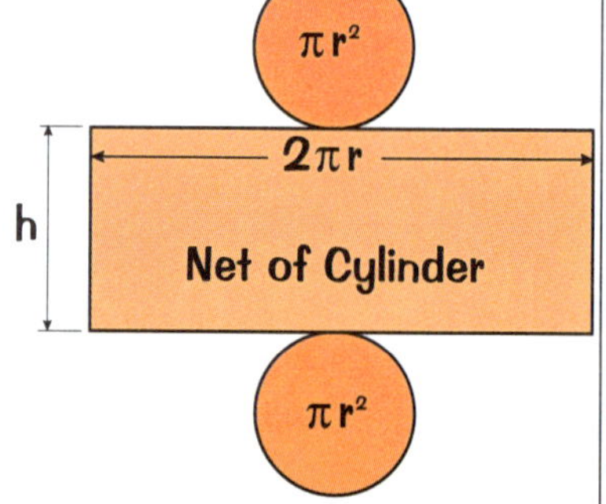

Note that the length of the rectangle is equal to the circumference of the circular ends.

When you see the words "lateral cylinder" and there's no $2\pi r^2$ in the formula, they're talking about the outside of a hollow tube.

The Acid Test: LEARN all the stuff on these two pages. Then turn over and try to write it all down. Keep trying until you can do it.

Find the surface area of these 2 shapes:

1)

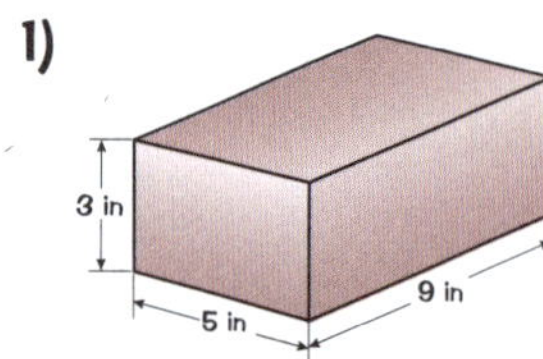

2)

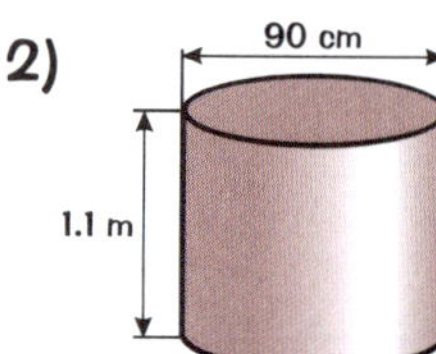

Estimating

1) Estimating

This is quite easy. To *estimate* something, this is all you do:

1) ROUND EVERYTHING OFF to nice easy CONVENIENT NUMBERS.
2) Then WORK OUT THE ANSWER using these nice easy numbers.

1) Don't worry about the answer being "wrong."
2) You're only trying to get a *rough idea* of the size of the proper answer.
In other words, is it about 20 or about 200, for example?

EXAMPLE "*Estimate the value of* $\frac{127 + 49}{56.5}$ *and show all your work.*"

Ans: $\frac{127 + 49}{56.5} \approx \frac{130 + 50}{60} \approx \frac{180}{60} \approx 3$ ("≈" means roughly equal to)

2) Estimating Areas and Volumes

This is easier than you might think too. All you have to do is this:

1) Draw or imagine a RECTANGLE OR BLOCK of similar size to the object.
2) Round off all lengths to the NEAREST WHOLE NUMBER, and work it out — easy.

1) Estimate the area of this shape:

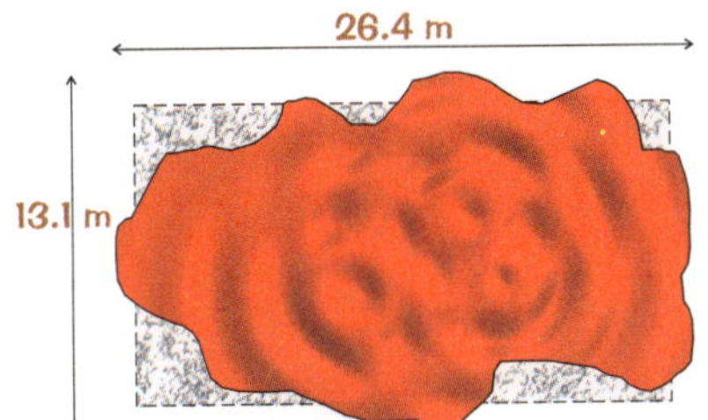

The area of the shape is *roughly the same* as the dotted rectangle shown:
i.e. 26 m × 13 m = $338\ m^2$

2) Estimate the volume of the bottle:

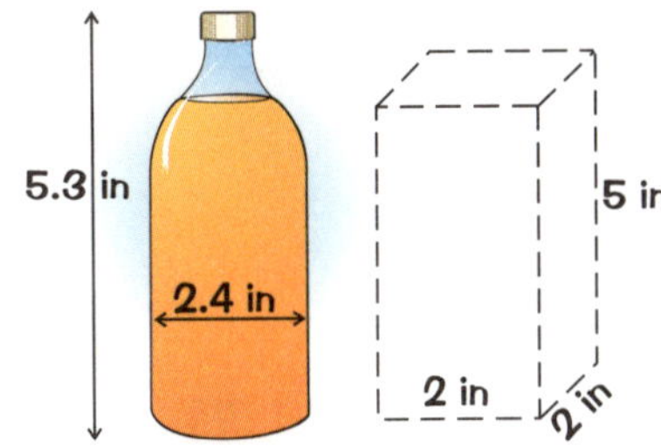

Volume of bottle is *approximately equal to* volume of rectangular block shown:
= 2 x 2 x 5 = $20\ in^3$

The Acid Test:

LEARN the 4 RULES for Estimating.
Then try these questions:

1) Without using your calculator, estimate the answer to this: $\frac{56.4 + 23.9}{36.1}$
2) Estimate the area of the palm of your hand in in^2.
3) Estimate the volume of a regular-sized can of beans in cm^3.
4) What is the approximate volume of your classroom, in yd^3?

Compute the length of the perimeter, the surface area of the faces, and the volume of a three-dimensional object built from rectangular solids. Understand that when the lengths of all dimensions are multiplied by a scale factor, the surface area is multiplied by the square of the scale factor and the volume is multiplied by the cube of the scale factor.
Relate the changes in measurement with a change of scale to the units used (e.g., in^2, ft^3) and to conversions between units ($1\ ft^2 = 144\ in^2$, $1\ in^3 \approx 16.38\ cm^3$).
Demonstrate an understanding of conditions that indicate two geometrical figures are congruent and what congruence means about the relationships between the sides and angles of the two figures.

Congruence and Enlargement

Congruent shapes

Congruence is another ridiculous math word that sounds really complicated when it's not. If two shapes are CONGRUENT, they're simply the same — the same size and the same shape.

So this just means the sides and angles are the same.

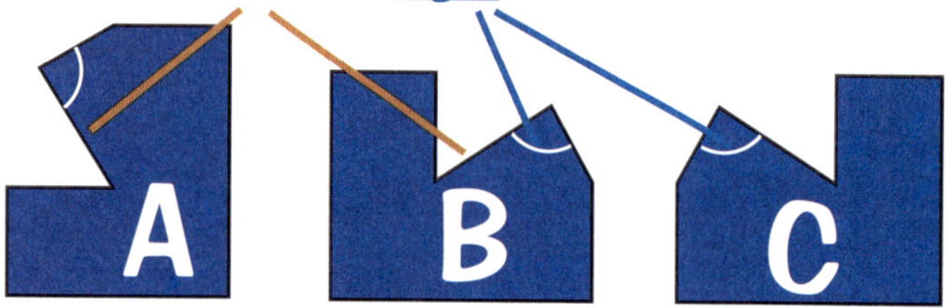

It's all pretty straightforward except for one trick:
Congruent shapes might be mirror images of each other (see B and C above).
So flipping a shape gives a shape congruent to the original.

Areas and Volumes of Enlargements

Ho ho! This little joker catches everybody out. The increase in area and volume is BIGGER than the scale factor. The rule is this simple:

For an enlargement of Scale Factor n:

The SIDES are	n	times bigger
The AREAS are	n^2	times bigger
The VOLUMES are	n^3	times bigger

Simple... but very forgettable.

FOR EXAMPLE, *if the Scale Factor is 2:*

1) The lengths are TWICE AS BIG, ($n = 2$)
2) Each area is 4 TIMES AS BIG, ($n^2 = 4$)
3) The volume is 8 TIMES AS BIG, ($n^3 = 8$)
as shown in the diagram:

All YOU have to do is REMEMBER it!

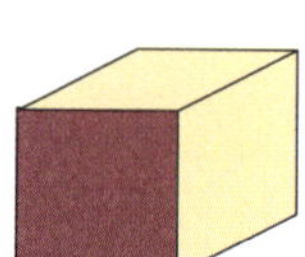

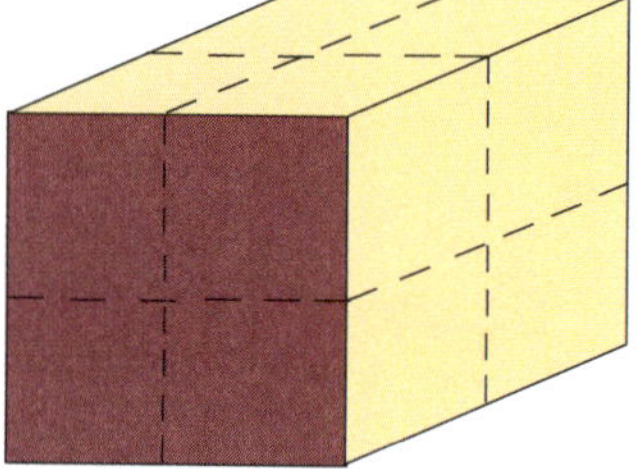

The Acid Test: LEARN exactly what "CONGRUENT" means, and the 3 Rules for INCREASE IN AREA AND VOLUME.

Now cover the page and write down what you've learned. Then REMEMBER it forever!

1) A soccer ball has a diameter six times bigger than a tennis ball's.
If the volume of the tennis ball is 180 cm^3, what's the volume of the soccer ball?

x- and y-Coordinates

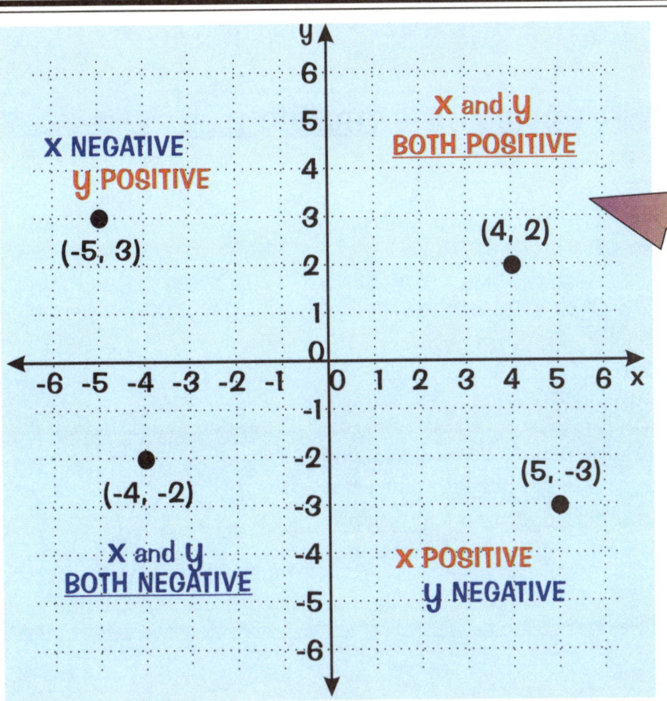

A graph has four different regions where the x- and y-coordinates are either positive or negative.

This is the easiest region by far because here ALL THE COORDINATES ARE POSITIVE.

You have to be *really careful* in the *OTHER REGIONS* though, because the x- and y-coordinates could be negative, and that always makes life much more difficult.

x-, y- Coordinates — get them in the Right Order

You must always give COORDINATES in parentheses, like this: (x, y)

(x, y)

And you always have to be really careful to get them *the right way around*, x first, then y. Here are *THREE POINTS* to help you remember:

1) The two coordinates are always in ALPHABETICAL ORDER, x then y.

2) x is always the flat axis going ACROSS the page.
In other words "x is a... cross" Get it — x is a "×"... (Hilarious isn't it.)

3) Remember it's always IN THE HOUSE (→) and then UP THE STAIRS (↑), so it's ALONG first, and then UP, or x-coordinate first, and then y.

The Acid Test:

LEARN the 3 Rules for getting x and y the right way around. Then turn over and write them down.

1) Write down the coordinates of the letters A to H on this graph:

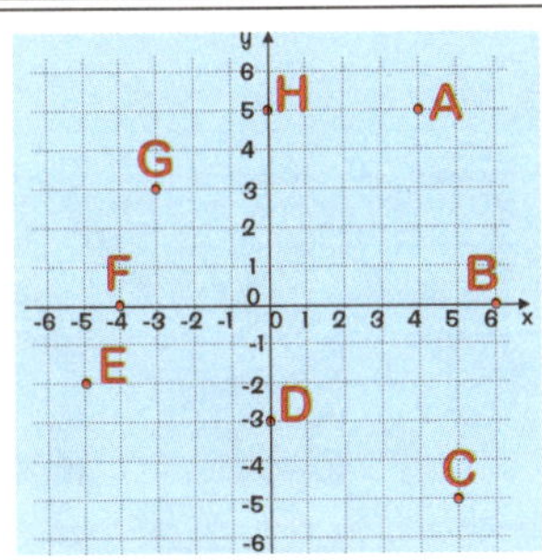

Graph Transformations

Translations

Translating an object on a graph just means **MOVING** it in a **STRAIGHT LINE**. There's no rotating, distorting, twisting, or weird stuff like that.

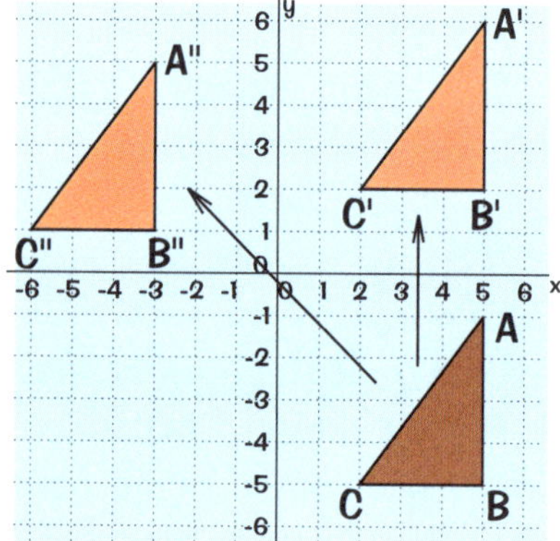

ABC to A'B'C' is a translation of 7 units up.

ABC to A"B"C" is a translation of 8 units left and 6 units up.

VECTORS

Translations can be described using vectors. The top number gives the horizontal change, and the bottom number gives the vertical change.

ABC to A'B'C' is a translation of $\begin{pmatrix} 0 \\ 7 \end{pmatrix}$

ABC to A"B"C" is a translation of $\begin{pmatrix} -8 \\ 6 \end{pmatrix}$

Reflections...

To reflect a shape, you need to know the **MIRROR LINE**. You get an **IMAGE** of the shape the other side of the mirror line, just like you'd get with a real mirror. **EACH POINT** of the shape is the **SAME DISTANCE BEHIND** the mirror line as the object is in front of it.

...in the x- or y-axis

The axis becomes the mirror line. You just need to make sure each point of the image is the same distance away from the axis as the real point.

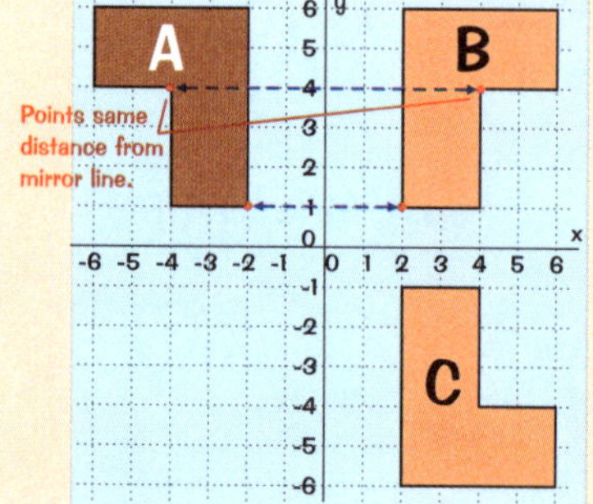

A to B is a reflection in the y-axis.

B to C is a reflection in the x-axis.

Reflecting in the x-axis is the same as reflecting in the line y = 0.
(And reflecting in the y-axis is the same as in the line x = 0.)

...in the line y = x

The other most common reflections are in the lines y = x and y = -x.

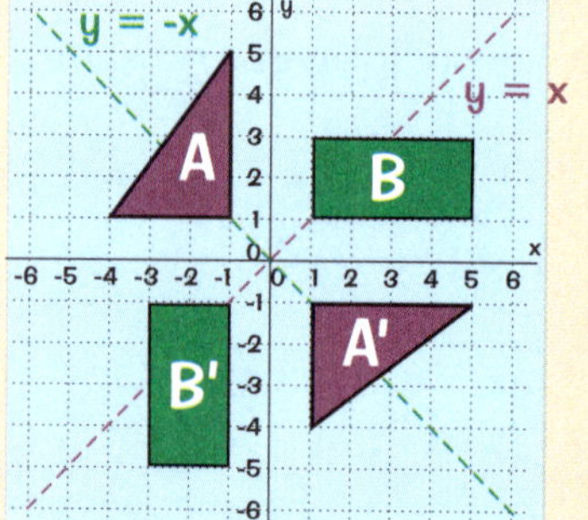

A to A' is a reflection in the line y = x

B to B' is a reflection in the line y = -x

The Acid Test: LEARN the names of the Two Transformations and the details that go with each.
When you think you know it, turn over and write it all down.

1) Describe, in full, these transformations: A ⟶ B, B ⟶ C, C ⟶ D.

Concept — Measurement and Geometry

Know and understand the Pythagorean theorem and its converse and use it to find the length of the missing side of a right triangle and the lengths of other line segments and, in some situations, empirically verify the Pythagoream theorem by direct measurement.

Pythagorean Theorem

1) THE PYTHAGOREAN THEOREM can only be used with RIGHT TRIANGLES.

2) If you know the lengths of any two sides, you can use it to find the third side.

Method:

The basic formula for the Pythagorean theorem is:

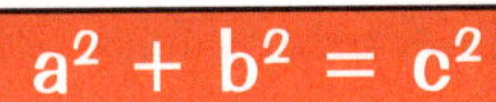

Remember that it can only be used on RIGHT TRIANGLES

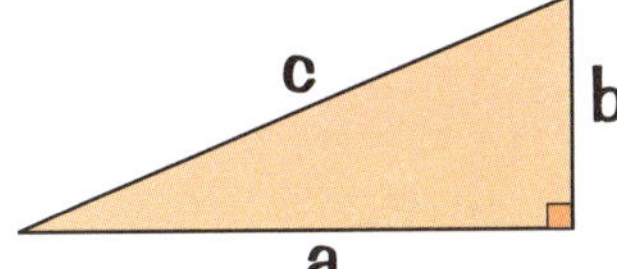

Don't forget that c is the long side — the *hypotenuse* — and that it's always the side *opposite* (or across from) *the right angle*.

The trouble is, the formula can be quite difficult to use. *Instead*, it's a lot better to *just remember* these *THREE SIMPLE STEPS*, which work every time:

1) Square Them

SQUARE THE TWO NUMBERS that you are given.

2) Add or Subtract

To find the *longest side*, ADD the two squared numbers.

To find *a shorter side*, SUBTRACT the smaller one from the larger.

3) Square Root

Once you have your answer, take the SQUARE ROOT.

Example 1: *"Find the missing side in the triangle shown."*

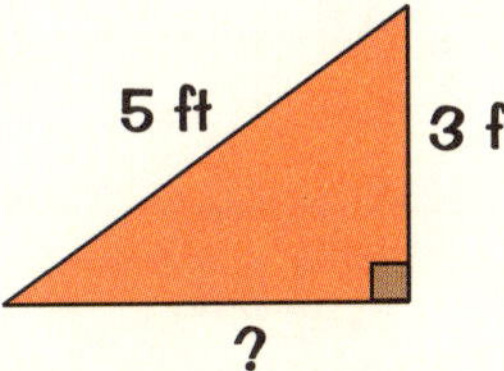

ANSWER: ❶ Square them: $5^2 = 25$, $3^2 = 9$

❷ You want to find a shorter side, so SUBTRACT: $25 - 9 = 16$

❸ Square root: $\sqrt{16} = 4$ So the missing side = 4 ft

You should always ask yourself: "Is it a *sensible answer*?" — in this case you can say "YES," because it's shorter than 5 ft (as it should be since 5 ft is the longest side), but not too much shorter.

Example 2: *"Find the length of the line segment shown."*

ANSWER: ❶ Work out how far across and up it is from A to B

❷ Treat this exactly like a normal triangle...

❸ Square them: $3^2 = 9$, $4^2 = 16$

❹ You want to find the longer side (the hypotenuse), so ADD: $9 + 16 = 25$

❺ Square root: $\sqrt{25} = 5$

So the length of the line segment = 5 units

For coordinates, see P.27

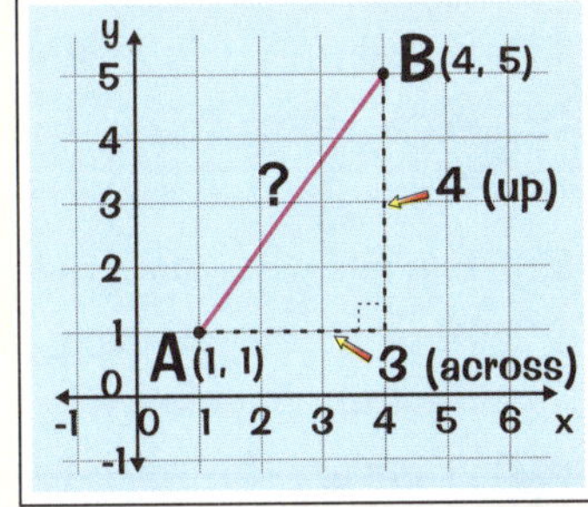

Here's one final thing to LEARN — you could easily get a question about it in the Exam:

The Pythagorean theorem works backwards too —
If you find a triangle where $a^2 + b^2$ is equal to c^2, then it MUST be a RIGHT TRIANGLE.

The Acid Test:

LEARN the 3 steps of the Pythagorean method.

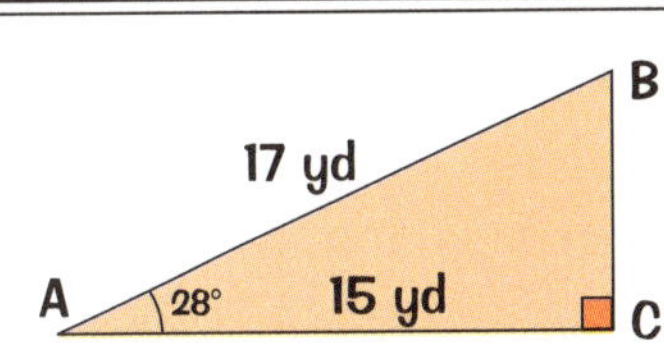

Now *turn over and write down what you've learned*.

1) Then apply the above method to find the missing side BC:

2) Another triangle has sides of 5 yd, 12 yd, and 13 yd. Is it a right triangle? How do you know?

Review Summary for Section Two

More difficult questions, *but just keep reminding yourself that they're the very best studying you can do*. These questions are very plain and very straightforward. They don't ask anything tricky, just whether or not you've actually *learned* all the *basic facts* in Section Two. It's really important to keep practicing these as often as you can.

Keep learning these basic facts until you know them

1) State an easy method for applying conversion factors.
2) Give 6 different conversions from one customary unit to another.
3) Give 8 different conversions from one metric unit to another.
4) Give 8 conversions between metric and customary units.
5) Write down the formulas for converting between Fahrenheit and Celsius.
6) What's the best way to convert units that have "2 parts" to them, e.g. km/h to mph?
7) A map is drawn to a scale of 1 cm to 4 km. Describe a method to figure out:
   a) the length on the map of an 8 km stretch of highway,
   b) the actual length of a road represented on the map by a 6 cm line.
8) What are the 2 steps for using a formula triangle?
9) What's the formula triangle for density? What's the formula triangle for speed?
10) What can you say about the units that come out of a formula?
11) Write down the Golden Rule. And the Silver Rule. Explain when you'd need to use each of these rules.
12) Write down the formulas for the areas and perimeters of five different kinds of shapes.
13) What is π? What are the two circle formulas? When do you use them?
14) Explain how to work out perimeters of complicated shapes.
15) What are the 3 steps for working out complicated areas?
16) What exactly is a prism?
17) Give the formulas for the volumes of a rectangular prism and another prism.
18) Explain what is meant by surface area and what a net is. How are the two related?
19) Write down the formulas for the surface area of a cube and the surface area of a cylinder.
20) State 2 rules for estimating an area or a volume.
21) Explain what congruence means.
22) If an object is enlarged by a scale factor of 3, how much bigger will the area and volume be for the new shape?
23) What are the rules for this in terms of a scale factor n?
24) Give 3 ways of remembering which way round the x- and y-coordinates go.
25) Sketch a pair of x- and y-axes and show where the x-coordinates and the y-coordinates are positive and negative in the four different regions.
26) Describe what "translating" an object means. Explain how translations can be described using vectors.
27) What's a mirror line? What's the relationship between an object, its reflected image, and the mirror line?
28) What is the formula for the Pythagorean theorem?
29) Describe an easy method for using the Pythagorean theorem.

Straight Lines You Should Just Know

You ought to know these simple graphs straight off with no hesitation:

1) Horizontal and Vertical lines: "x = a" and "y = b"

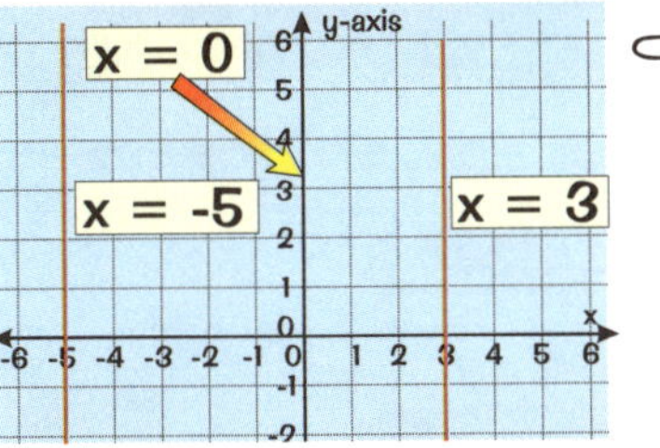

x = a is a vertical line through "a" on the x-axis.

y = b is a horizontal line through "b" on the y-axis.

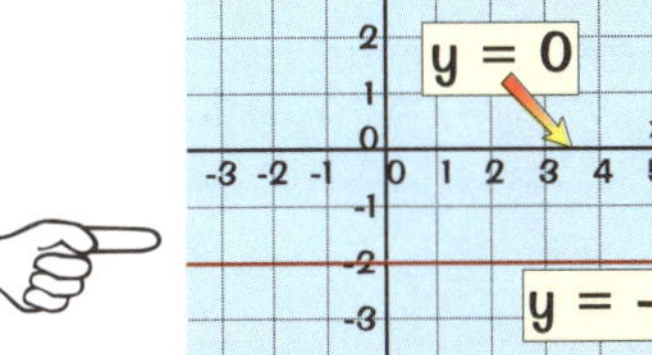

Don't forget: the y-axis is also the line x = 0

Don't forget: the x-axis is also the line y = 0

2) The Main Diagonals: "y = x" and "y = -x"

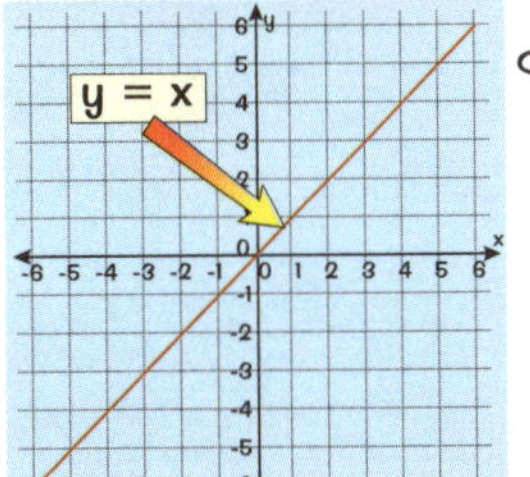

y = x is the main diagonal that goes UPHILL from left to right.

y = -x is the main diagonal that goes DOWNHILL from left to right.

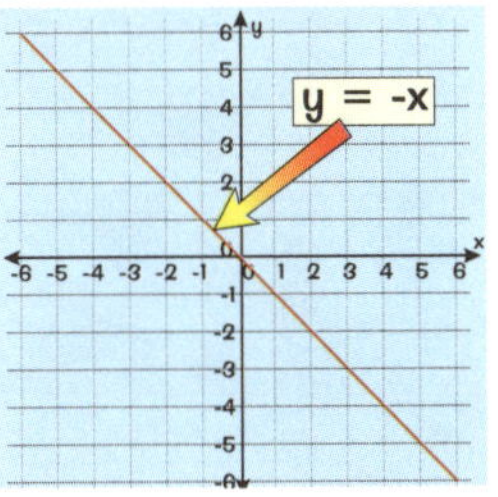

3) Other Sloping Lines Through the Origin: "y = mx" and "y = -mx"

y = mx and y = -mx are the equations for A SLOPING LINE THROUGH THE ORIGIN.

The value of "*m*" is the *SLOPE* of the line — the BIGGER the number the STEEPER the slope, and a MINUS SIGN tells you it slopes DOWNHILL as shown by the ones here:

The slope of a horizontal line (e.g. y = 3) is 0.
The slope of a vertical line (e.g. x = 2) is ∞ (infinity — ooh).

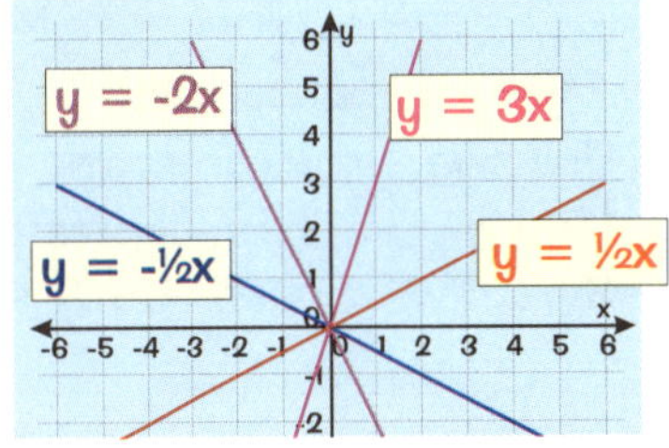

All Other Straight Lines

Other straight-line equations are a little more complicated, and pages 34-35 show three methods for drawing them. The first step, of course, is identifying them in the first place.
Remember:

(Either of these "somethings" could be zero. Actually, so could the number.)

All straight line equations just contain "*something x, something y, and a number*."

Straight lines:

$x - y = 0$ | $y = 2 + 3x$
$4x = 7$ | $4x - 3 = 5y$
$3y + 3x = 12$ | $6y = 0$
$5(x + 3y) = 5$ | $12x - 7 = 3(x + 4y)$

NOT straight lines:

$y = x^3 + 3$ | $2y - \frac{1}{x} = 7$
$\frac{1}{y} + \frac{1}{x} = 2$ | $x(3 - 2y) = 3$
$x^2 = 4 - y$ | $xy + 3 = 0$
$2x + 3y = xy$ | $y = \frac{1}{2}\sin x$

The Acid Test:

LEARN all the specific graphs on this page and also how to identify straight line equations.

Now turn over the page and write down everything you've learned.

Graph linear functions, noting that the vertical change (change in y-value) per unit of horizontal change (change in x-value) is always the same and know that the ratio ("rise over run") is called the slope of a graph.

Slope of a Straight Line

Figuring out the slope of a straight line is a slightly involved business, and there are quite a few things that can go wrong.

Once again though, if you *learn and follow the steps below* and treat it as a STRICT METHOD, you'll have a lot more success than if you try to fudge your way through it like people usually do.

Strict Method for Finding the Slope

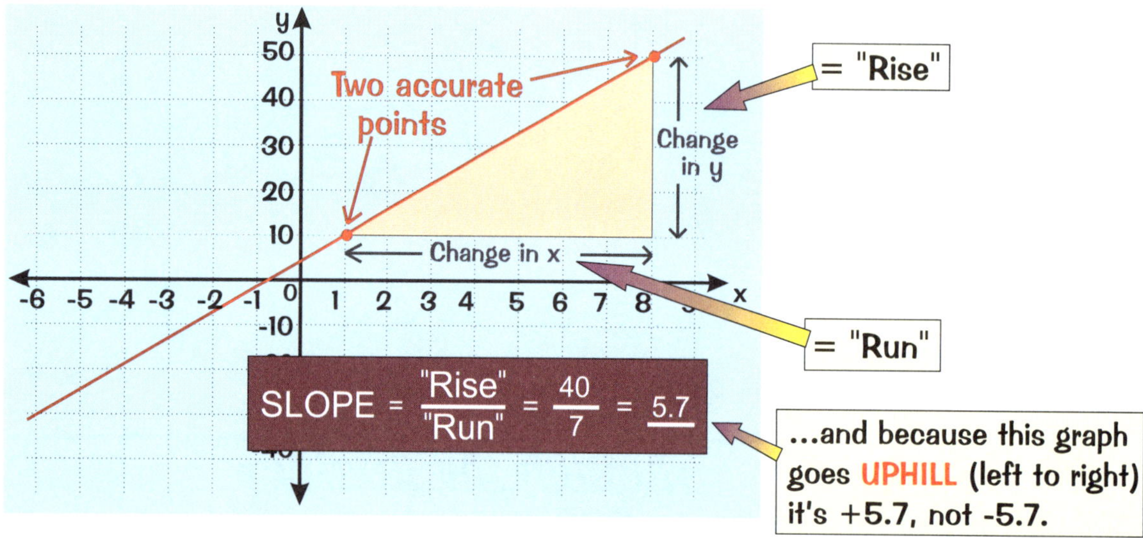

1) Find *TWO ACCURATE POINTS*, reasonably far apart.

Use two values from the *upper right quadrant* if possible.
(To keep all the numbers positive and so reduce the chance of errors.)
Here, I've chosen (1, 10) and (8, 50) because the coordinates are all whole numbers.

2) *COMPLETE THE TRIANGLE* as shown.

3) Find the *CHANGE IN y* and the *CHANGE IN X*.

Make sure you do this *using the SCALES on the x- and y-axes,* not by measuring it with a ruler.

(So in the example, the change in x is 8 – 1 = 7 — this is NOT 7 in, 7 cm, or 7 ft, it's "7 units along the x-axis.")

4) *LEARN* this formula, and use it:

$$\text{SLOPE} = \frac{\text{RISE}}{\text{RUN}}$$

Make sure you get it the right way up too. Remember, RISE is the VERTICAL change and RUN is the HORIZONTAL change, and the slope is VERy HOt — VERtical over HOrizontal.

5) Finally, is the slope *POSITIVE* or *NEGATIVE*?

If it slopes UPHILL left → right (↗) then it's +.
If it slopes DOWNHILL left → right (↘) then it's – (so put a minus (–) in front of it).

Slope of a Straight Line

What the Slope of a Graph MEANS

No matter what the graph, THE MEANING OF THE SLOPE is always simply:

(y-axis UNITS) PER (x-axis UNITS)

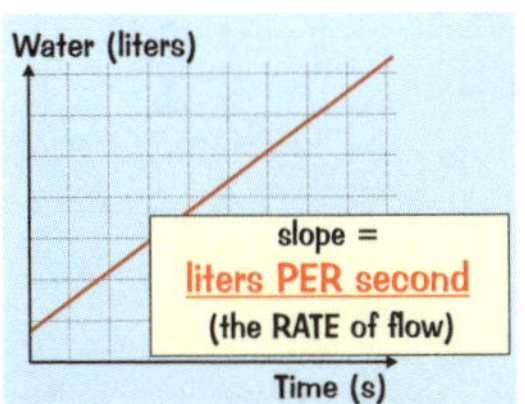

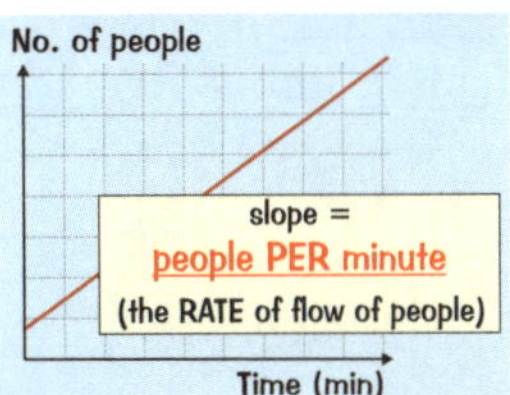

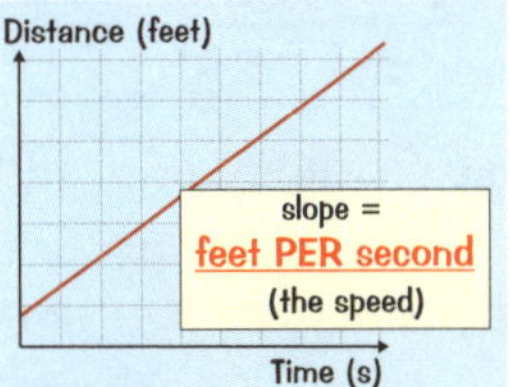

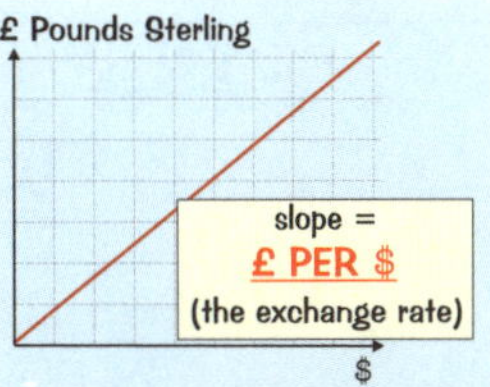

Some slopes have special names like exchange rate, speed, or some other kind of rate, but once you've written down "something PER something" using the y-axis and x-axis UNITS, it's then pretty easy to work out what the slope represents.

A Typical Slope Question

"For each of the graphs below, find the heart rate represented by the line."

a)

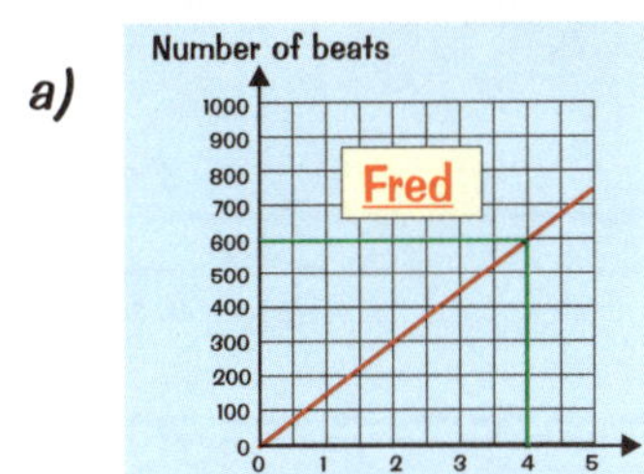

b)

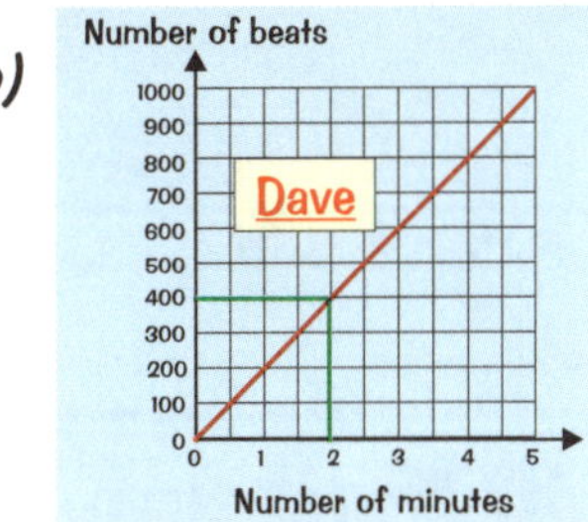

c)

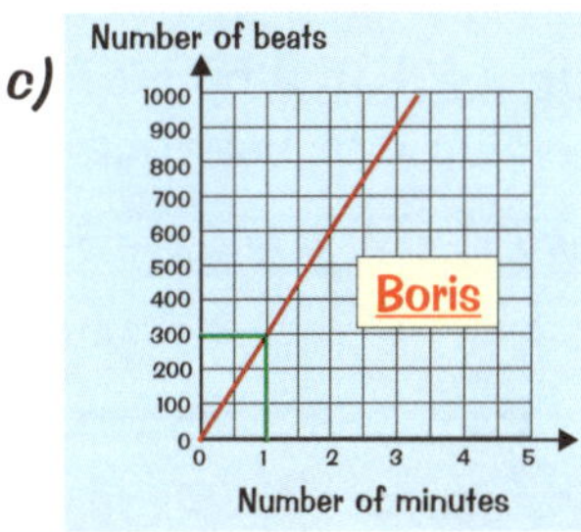

ANSWER:

Just work out the slope of each graph like on the opposite page...

a) Slope = $\frac{600-0}{4-0}$ = 150 beats per minute

b) Slope = $\frac{400-0}{2-0}$ = 200 beats per minute

c) Slope = $\frac{300-0}{1-0}$ = 300 beats per minute

If your graph goes through the origin, use it as one of the two points you choose... ...it makes life a little easier.

Whatever you do, don't forget the units. They're mega-important.

(If you're thinking these heart rates are ridiculous, don't worry. Fred, Dave, and Boris are my pet hedgehog, ferret, and rat.)

The Acid Test:

LEARN the FIVE STEPS for finding the slope, then turn over and WRITE THEM DOWN from memory.

1) Plot these 3 points on a graph:
(0, 3) (2, 0) (5, -4.5)
and then connect them with a straight line.
Now carefully apply the FIVE STEPS to find the slope of the line.

2) This graph (possibly the most exciting one you'll ever see) shows the amount of gas being dispensed from a gas pump.

By working out the slope, find the RATE represented on this graph.

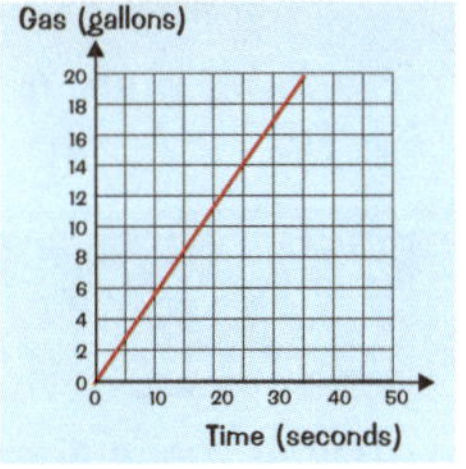

Concept — Algebra and Functions
Graph linear functions, noting that the vertical change (change in y-value) per unit of horizontal change (change in x-value) is always the same and know that the ratio ("rise over run") is called the slope of a graph.
Algebra 1 Students graph a linear equation and compute the x- and y-intercepts.

Graphing Straight Lines

Some people wouldn't know a straight line equation if it ran up and bit them, but they're pretty easy to spot really — they'll have one or two letters (usually x and y), but nothing fancy like squares or cubes (see the examples at the bottom of P.31).

Anyway — if you can draw the graphs of straight line equations, it'll come in handy for Exam questions. "y = mx + b" is the hard way of doing it (see P.35), but here's TWO NICE EASY WAYS:

1) The "Table of 3 Values" Method

You can EASILY draw the graph of ANY STRAIGHT LINE using this EASY method:

1) Choose 3 values of x and draw up a little table,
2) Work out the y-values,
3) Plot the coordinates, and draw the line.

If it's a *straight line equation*, the 3 points will be in a *dead straight line*, which is *the usual check* you do when you've drawn it — *if they aren't*, then it could be a *curve* and you'll need to do *more values in your table* to find out what on earth's going on (see P.38).

Example: *"Draw the graph of y = 2x – 3"*

1) DRAW UP A TABLE with some *suitable values* of x. Choosing x = 0, 2, 4 is usually good enough — i.e.

x	0	2	4
y			

2) FIND THE y-VALUES by putting each x-value into the equation:

x	0	2	4
y	-3	1	5

(e.g. When x = 4, y = 2x – 3 = (2 × 4) – 3 = 5)

3) PLOT THE POINTS and DRAW THE LINE.

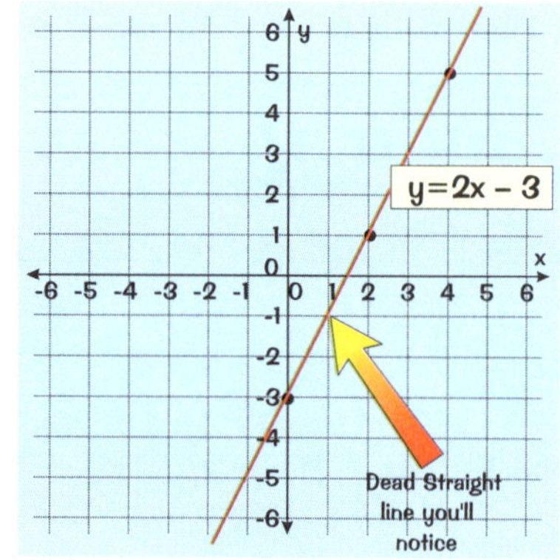

2) The "x = 0, y = 0" Method

This is especially good for equations of the form: "ax + by = c"

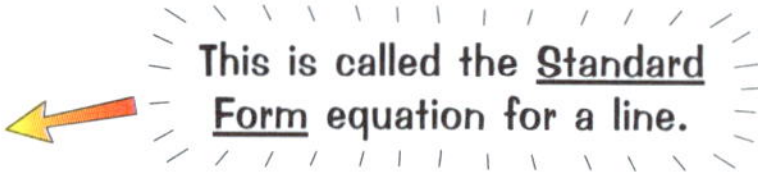

1) Set x = 0 in the equation and find y — this is the y-intercept (i.e. where it crosses the y-axis).
2) Set y = 0 in the equation and find x — this is the x-intercept (i.e. where it crosses the x-axis).
3) Plot these two points and connect them with a straight line — *and just hope it should be a straight line, since with only 2 points you can't really tell, can you!*

Example: *"Draw the graph of 5x + 3y = 15."*

1) Putting x = 0 gives "3y = 15" ⇒ y = 5
2) Putting y = 0 gives "5x = 15" ⇒ x = 3
3) So plot y = 5 on the y-axis and x = 3 on the x-axis and connect them with a straight line:

Only doing 2 points is risky unless you're sure the equation is definitely a straight line — but then that's the big thrill of living life on the edge, isn't it.

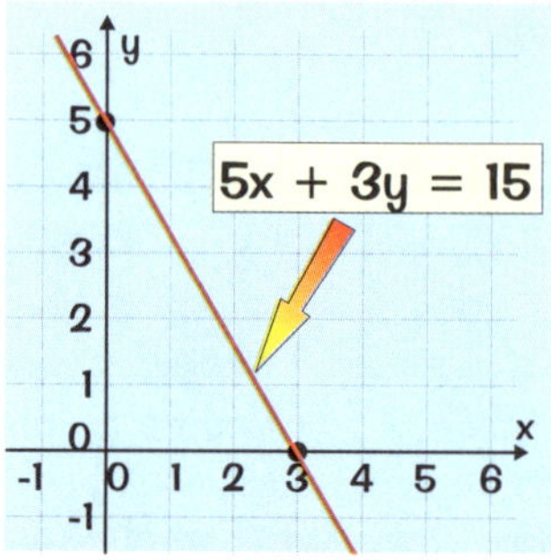

The Acid Test:

LEARN the details of these TWO EASY METHODS, then turn over and write down all you know.

1) Graph these equations using both methods: a) y = 4 + x b) 4y + 3x = 12 c) y = 6 – 2x

Graphing Straight Lines

"y = mx + b" is called the slope-intercept form of a straight line because you can magically read off the slope and the point where it hits the y-axis (the y-intercept).

3) i) Getting the equation into "y = mx + b" form

The first thing you need to be able to do is rearrange equations into "y = mx + b" like this:

Straight line:		Rearranged into "y = mx +b"	
$y = 2 + 3x$	→	$y = 3x + 2$	$(m = 3, b = 2)$
$2y - 4x = 7$	→	$y = 2x + 3\frac{1}{2}$	$(m = 2, b = 3\frac{1}{2})$
$x - y = 0$	→	$y = x + 0$	$(m = 1, b = 0)$
$4x - 3 = 5y$	→	$y = 0.8x - 0.6$	$(m = 0.8, b = -0.6)$
$3y + 3x = 12$	→	$y = -x + 4$	$(m = -1, b = 4)$

REMEMBER: "m" equals the slope of the line.
"b" is the "y-intercept" (where the graph hits the y-axis).

BUT WATCH OUT: people mix up "m" and "b" when they get something like $y = 5 + 2x$.
Remember, "m" is the number IN FRONT OF THE "x" and "b" is the number ON ITS OWN.

3) ii) Drawing a Straight Line using "y = mx + b"

Method:

1) Get the equation into the form "y = mx + b."
2) IDENTIFY "m" and "b" CAREFULLY.
3) PUT A DOT ON THE y-AXIS at the value of b.
4) Then go ALONG ONE UNIT and up or down by the value of m and make another dot.
5) Repeat the same "step" a few times in both directions, then draw a line through the dots:
6) Finally, CHECK that the slope LOOKS RIGHT.

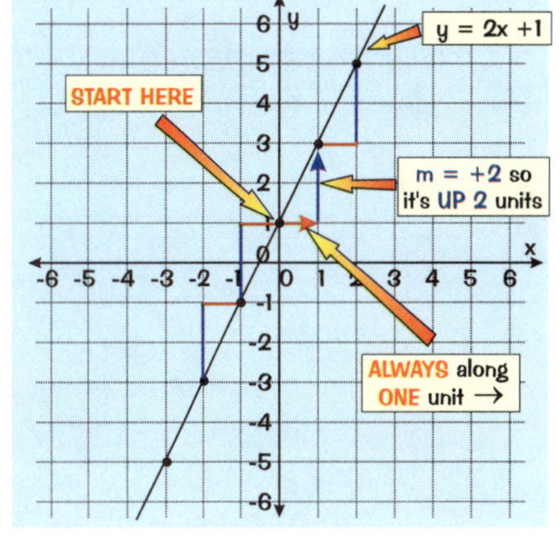

This graph shows the process for the equation "y = 2x + 1":

1) "b" = 1, so put a first dot at y = 1 on the y-axis.
2) Go along 1 →, and then up by 2 ↑, because "m" = +2.
3) Repeat the same step, 1 → 2 ↑, in both directions. (i.e. 1 ← 2 ↓ the other way)
4) CHECK: a slope of +2 should be quite steep and uphill left to right — which it is.

***IMPORTANT:** If "m" isn't a whole number, you'll have to go up or down a fraction of a unit in step 4. It's dead easy to go wrong, so always check and double-check your answer.*

The Acid Test:

LEARN all the details of y = mx + b equations, and make sure you know how to graph them.

1) Write these equations in the form y = mx + b and state the value of m and b in each case:
   a) $2x + y = 3$ b) $4x + 2y = 1$
2) Using "y = mx + b," draw the graphs of $y = x - 3$ and $y = 4 - 2x$.

Typical Questions on Straight Lines

Oh look, here's a double page all about the types of questions you'll get in the Exam. Not only is it awesomely exciting, but it looks pretty important too.

Find the x- or y-intercept

"Find the x-intercept of the line $5y + 3x = -3$."

This x-intercept is just where the line crosses the x-axis (see P.34).
At any point on the x-axis, $y = 0$ (it's obvious when you look at a graph).
Sticking $y = 0$ in the equation gives $0 + 3x = -3$, which gives $x = -1$.
So the coordinates of the x-intercept are $(-1, 0)$.

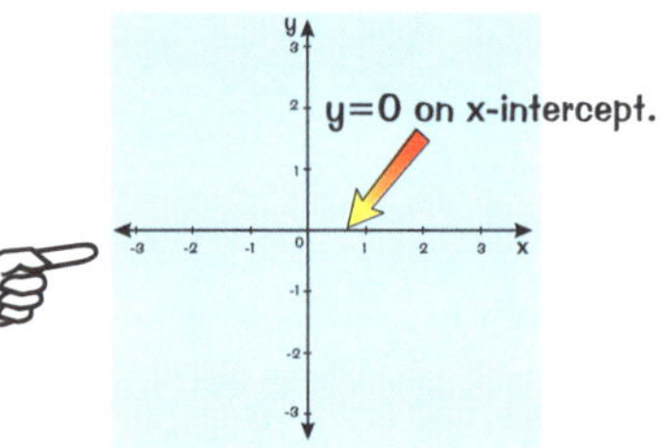

Find Out Whether a Point is On the Line

"Determine whether the point (6, 11) lies on the line $2y = 3x + 4$."

YOU DON'T EVEN NEED THE GRAPH — if it's on the line, it'll fit the equation.

Simply stick $x = 6$ and $y = 11$ in the equation and see if it works:
The equation is: $2y = 3x + 4$ — so you get: $(2 \times 11) = (3 \times 6) + 4$
So $22 = 18 + 4$ — yep, that's right. The point (6, 11) does indeed lie on the line $2y = 3x + 4$.

Getting the Equation...

1) From a Picture

"Find the equation of the graph shown on the right."

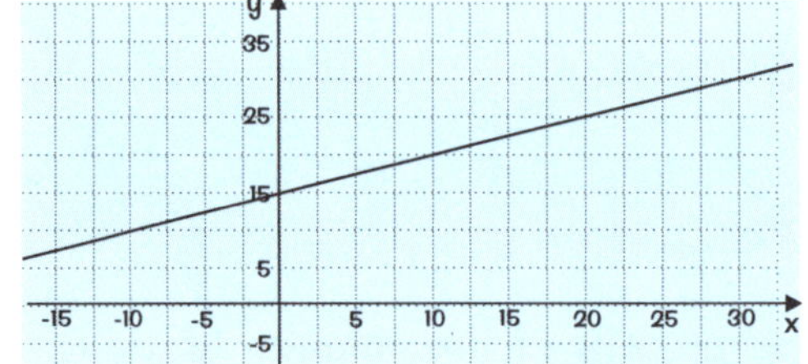

FOLLOW THIS EASY METHOD:

1) Find the y-intercept. This is the value of "b."
2) Find the value of the SLOPE (see P.32). This is the value of "m."
3) Now just put these values for "m" and "b" into "$y = mx + b$" — and there you have it!

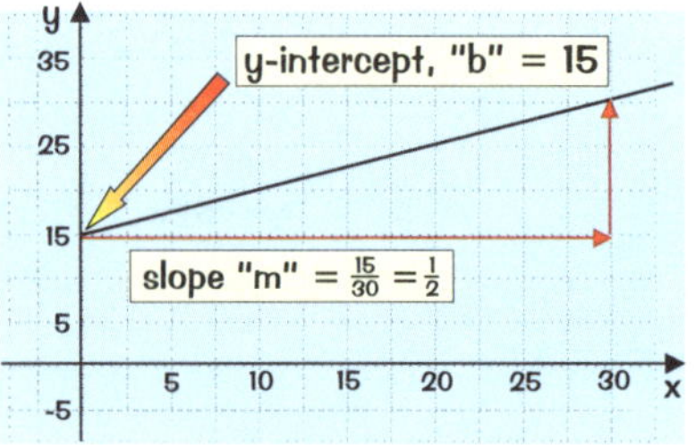

So in this case, $m = \frac{1}{2}$ and $b = 15$, so "$y = mx + b$" becomes "$y = \frac{1}{2}x + 15$."

2) From Pieces of Information

"Find the equation of the straight line passing through the point (6, 2) with a slope of $-\frac{1}{3}$."

It's really simple: Just use $y = mx + b$.

The value of m is $-\frac{1}{3}$, which means the equation is "$y = -\frac{1}{3}x + b$." So you just need to find the value of b.
To do this, stick the x- and y-values of the point you know into the equation.

$y = 2$ when $x = 6$ $\Rightarrow 2 = -\frac{1}{3} \times 6 + b \Rightarrow 2 = -2 + b \Rightarrow b = 4$

So putting it all together gives "$y = -\frac{1}{3}x + 4$."

Typical Questions on Straight Lines

Questions about parallel lines and translating lines are incredibly easy, if you know these basic facts:

Parallel Lines

PARALLEL LINES have the SAME SLOPE,
(but different y- and x-intercepts)

...and that's all you need to know.

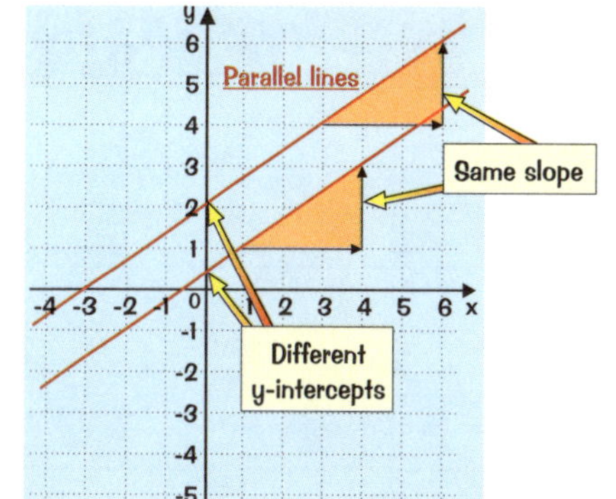

So you can tell two lines are parallel by looking at $y = mx + b$:

$y = 3x + 5$ $y = 3x - 1$ — PARALLEL ($m = 3$)

$y = 5x + 5$ — ODD ONE OUT ($m = 5$)

Moving (Translating) Lines

Moving a graph vertically changes the y-intercept of the graph (i.e. changes b in $y = mx + b$).
You end up with a line parallel to the original one — the slope hasn't changed.

TRANSLATING a line up or down changes the value of b.
The SLOPE DOESN'T CHANGE.

"translating" is just a fancy word for "moving"

A translation up of 5 increases b by 5.
A translation down of 7 subtracts 7 from the value of b.

See P.28 for more on transformations.

Example 1:

"What is the equation of the line parallel to $y = \frac{1}{2}x + 3$ that intercepts the y-axis at $y = -5$?"

Remember $y = mx + b$: The line is parallel, so m stays the same ($m = \frac{1}{2}$).
And b is always the y-intercept, which here is -5. So the answer is $y = \frac{1}{2}x - 5$.

Example 2:

"Find the new equation of the line $y = 2x + 2$ after it has been translated down 5 units."

All you do is take 5 from "b" to give -3. So the answer is "$y = 2x - 3$."

See — I told you they were easy, but you just didn't believe...

The Acid Test:

These examples will prepare you really well for the questions you'll get in the Exam, so LEARN THEM ALL.

1) Find the x- and y-intercepts of the line $5x + 3y = 3$.
2) Using "$y = mx + b$" find the equations of these 3 graphs →
3) Find the equation of the straight line passing through the point (-3, 10) with a slope of 4.
4) Find the equation of the line from 3) after it's been translated up 37 units.
5) Which 2 of these lines are parallel? $y = 2x - 3$ $y = 3x - 3$ $y - 2x = 5$

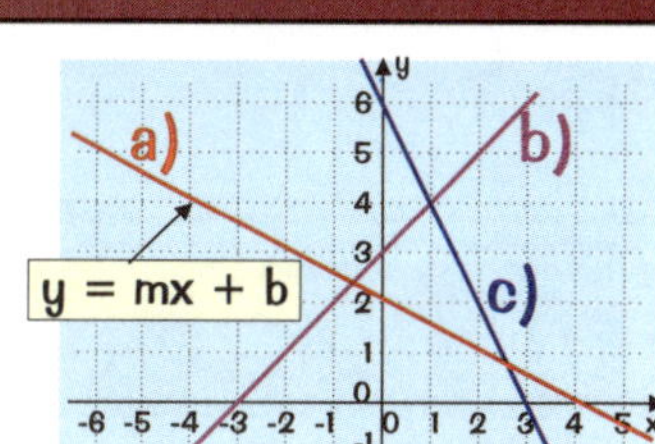

Curves You Should Recognize

There are two types of curves that you should know the basic shape of just from looking at their equations — it really isn't as difficult as it sounds.

$y = x^2$ — "u" shaped parabola

"U" shaped graphs like these are called parabolas.

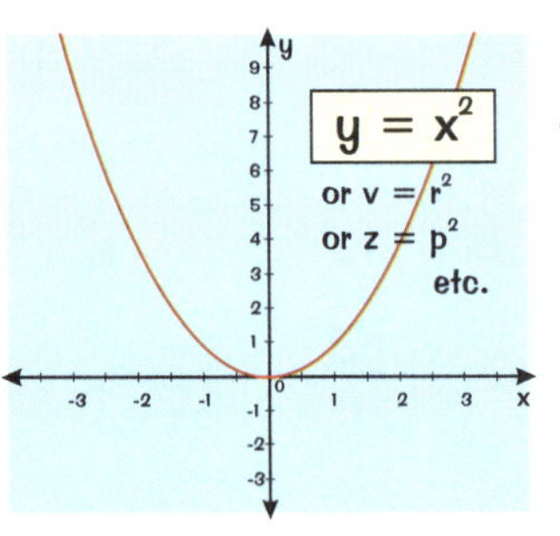

$y = x^2$ is a u-shaped graph that passes through the origin and is symmetrical about the y-axis.

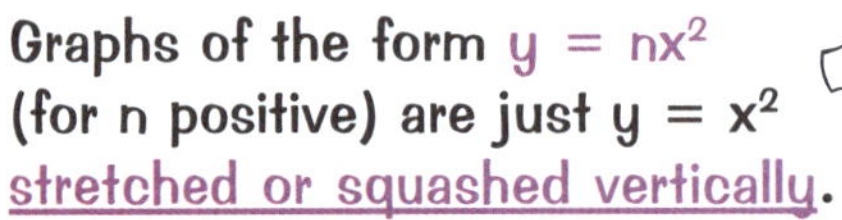

Graphs of the form $y = nx^2$ (for n positive) are just $y = x^2$ stretched or squashed vertically.

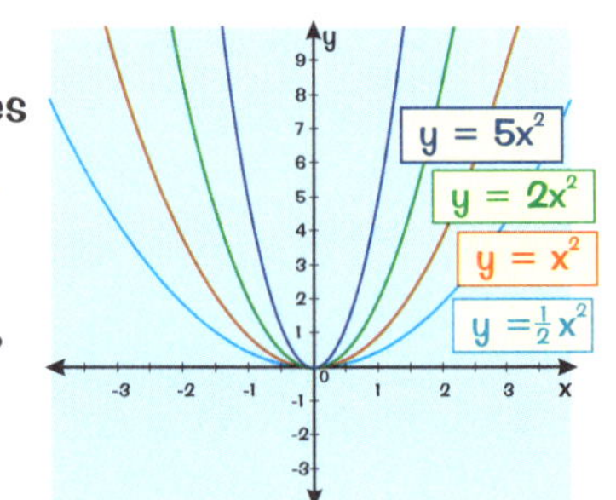

$y = -x^2$ — "n" shaped parabola

Putting a minus sign in front flips the graph upside down about the x-axis, making it n-shaped.

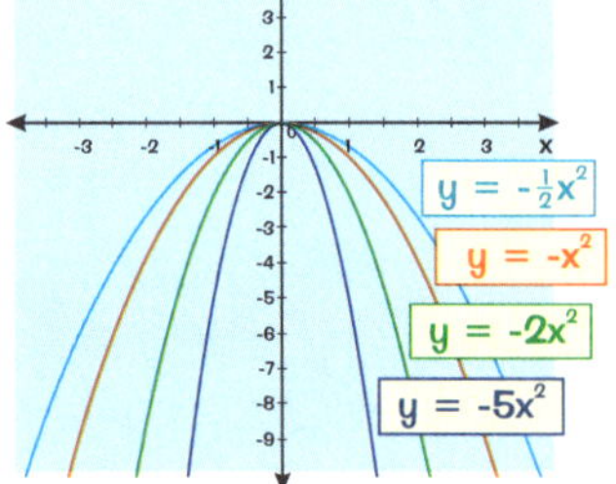

$y = x^3$ — corner to corner and flat in the middle

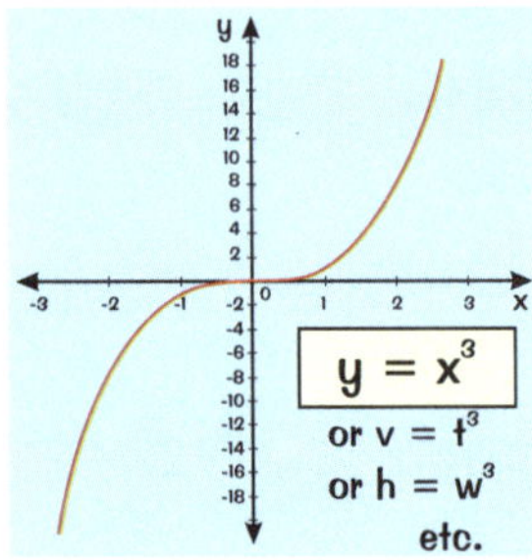

$y = x^3$ goes upwards from bottom left to top right, flattening out as it passes through the origin.

Graphs like $y = 2x^3$, $y = 5x^3$, $y = nx^3$ (for n positive) are just the same except stretched or squashed vertically.

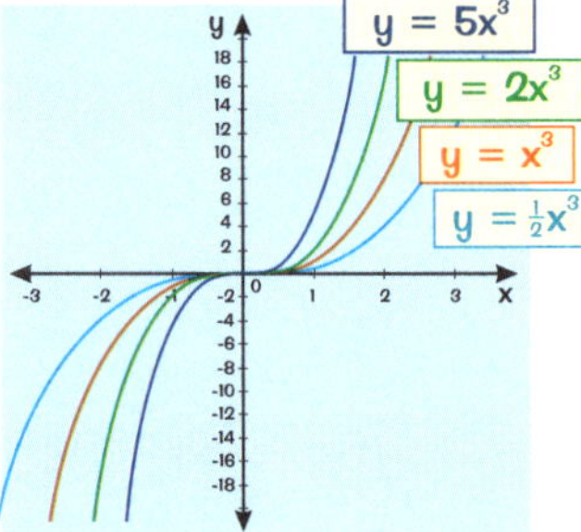

Just like x^2 graphs, sticking a minus sign in front of the x^3 flips the graph over about the x-axis.

So it goes down from top left to bottom right.

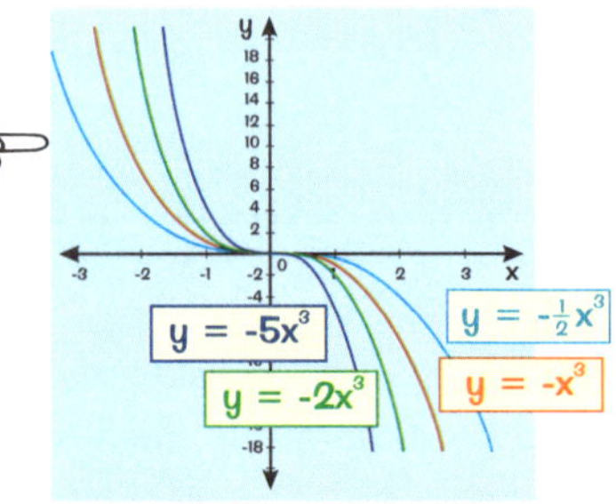

The Acid Test: LEARN the 2 Types of Graphs and the effect of putting a number or a minus sign in front. Then turn over and sketch examples of each graph.

1) Sketch these equations on the same axes: a) $y = 3x^2$ b) $y = -x^2$
2) Now do the same with these ones: a) $y = \frac{1}{2}x^3$ b) $y = -x^3$

Getting Answers from Your Graph

Reading graphs is so easy it's practically free points in the Exam. You'd be a fool not to learn it.

Find the Value and draw a line to the Graph

1) FOR A SINGLE CURVE OR LINE, you ALWAYS get the answer by *drawing a straight line to the graph from one axis*, and then *down or across to the other axis*, as shown here:

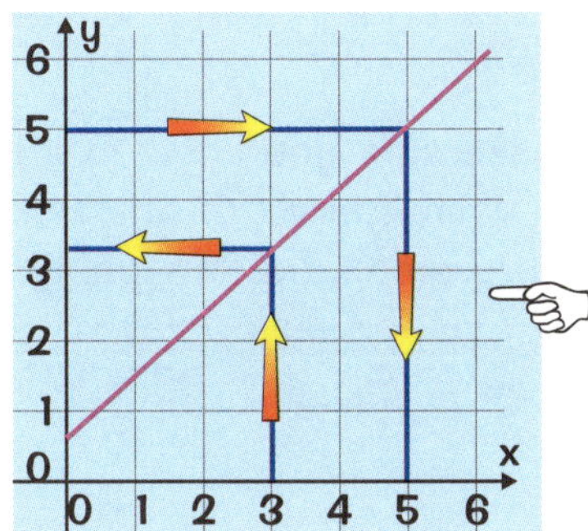

You should be *expecting* this so that even if you don't understand the question, you can still have a pretty good stab at it.

If the question says *"Find the value of y when x is equal to 3,"* ALL YOU DO IS THIS: start at 3 on the x-axis, go straight up to the graph, then straight over to the y-axis and read off the value, which in this case is *y = 3.3*.

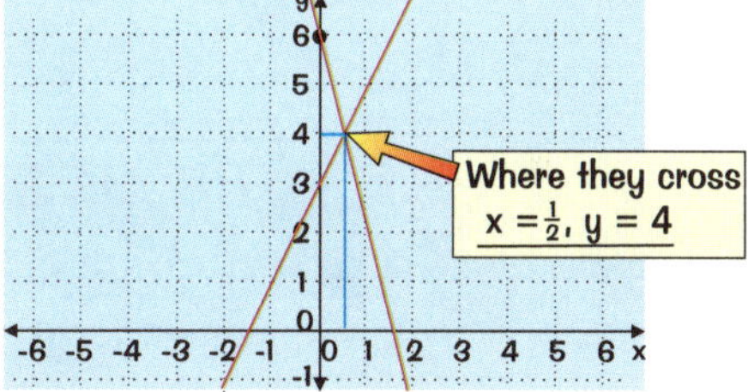

2) IF TWO LINES CROSS...
you can bet your very last cupcake the answer to one of the questions will simply be:
THE VALUES OF x AND y WHERE THEY CROSS
and you should be expecting that *before they even ask it!*

Example:

A small piece of moldy cheese is fired from a catapult. The height, h, of the cheese at a time, t, is shown on the graph below. Using a graphing method find:
a) the times at which the moldy cheese is at a height of 25 m, and b) its height after 2½ s.

ANSWER:

a) Height is on the vertical axis. Find 25 meters on this axis and draw a line across the graph. At the two points where the line hits the graph, draw a vertical line down to the other axis and read off the values:
t = 1.4 s and t = 3.6 s.

b) To get the height after 2½ seconds, find 2½ on the horizontal axis and draw a line up to the graph. Where the line hits the graph, draw a horizontal line back to the vertical axis and read the value, which is h = 31.5 m.

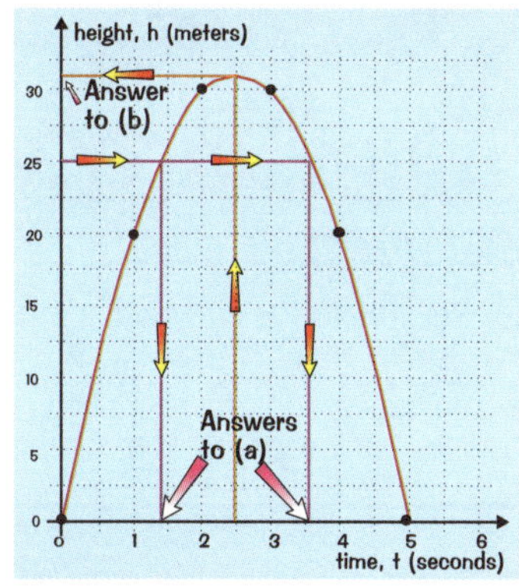

The Acid Test:

LEARN EVERYTHING on this page. In particular, make sure you understand the importance of the cheese being moldy in the example.

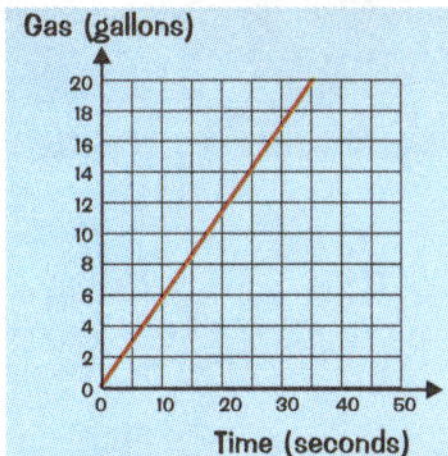

1) The graph (left) is the most exciting graph in the world (from P.33).
Use this graph to find: a) how much gas is pumped in 30 s.
b) how long it takes to pump 15 gallons.

2) Using the moldy cheese graph above,
a) find the times where the cheese is at a height of 15 m.
b) find the length of time for which the cheese was above 15 m.

Travel Graphs

Travel graphs are great. You'll just love them. I promise...

Travel Graphs — Always the same and always easy

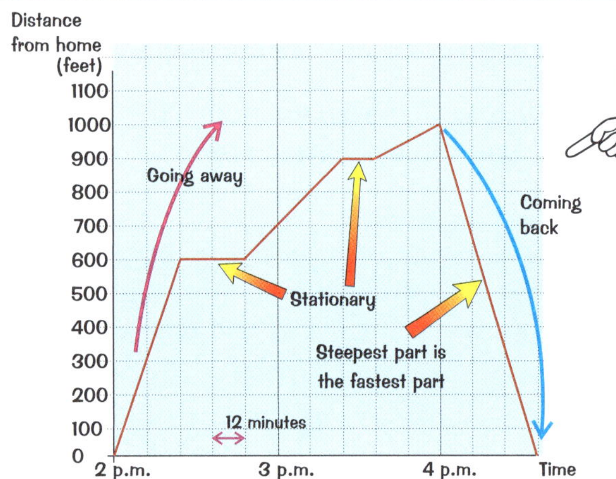

1) Make sure you know *all these details* about travel graphs.

2) Also, always make sure you know exactly *how much time* each interval on the time axis actually represents. On this graph there are *5 divisions* for each *hour*, so each one must be *12 minutes* (60 ÷ 5).

THE SIX KEY POINTS ABOUT TRAVEL GRAPHS

1) A TRAVEL GRAPH is always DISTANCE (↑) against TIME (→).
2) For any section, SLOPE = SPEED, but watch out for the UNITS.
3) FLAT SECTIONS are where it's STOPPED.
4) The STEEPER the graph, the FASTER the thing's going.
5) UPHILL SECTIONS mean it's TRAVELING AWAY from its starting point.
6) DOWNHILL SECTIONS mean it's COMING BACK toward its starting point.

You only get downhill sections on graphs showing the "distance from" somewhere. A lot of graphs just show the total distance traveled, which means they'll only go up.

A Typical Tricky Question:

"What's the average speed of the journey from 3 p.m. to 4 p.m. (in feet per minute)?"

This looks a little weird because the speed isn't constant between 3 p.m. and 4 p.m. But it's easy. To find the AVERAGE SPEED, just pretend the graph is a straight line between the two times, and find the slope.

Speed = slope
= 300 ft/60 min
= 5 ft/min (feet per minute)

You have to use minutes here to get the answer in "feet per minute."

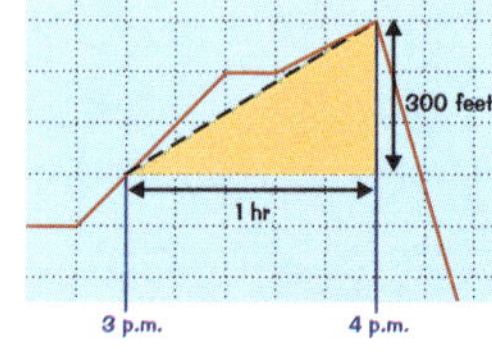

Never forget the units when writing down a speed — they're insanely important.

The Acid Test:

LEARN all the details on the graph above and then the Six Key Points for Travel Graphs.

Now cover the page and write down everything you've learned.

1) For the travel graph above, figure out the average speed from 2 p.m. to 3 p.m. in miles per hour. Remember, 3 feet = 1 yard, and 1 mile = 1760 yards. (See — I'm nice really. ☺)
2) Also, describe the whole sequence of events between 2 p.m. and 4:36 p.m.

Review Summary for Section Three

Aren't graphs great. I think everyone likes graphs. Anyway, here are the scary questions to find out what you know. By the way, I hope you haven't started trying to kid yourself that these aren't proper math questions so you don't have to bother with them. Oh no. Math is _packed_ with _facts_ — and you need to know them _to be able to do it_. All these questions do is see how many of the simple facts you've learned so far. Try it now and scare yourself. Then learn some stuff and try again.

Keep learning these basic facts until it kills you

1) What type of line is $x = a$?
2) What type of line is $y = b$?
3) What type of line is $y = mx$?
4) What's special about $y = x$ and $y = -x$?
5) What does the equation of a straight line look like?
6) What makes them different from equations that aren't straight lines?
7) What is the formula for the slope of a line? Describe an easy way to remember it.
8) Describe a method for finding the slope of a line.
9) How do you determine the *meaning* of the slope of a graph?
10) What x- and y-axes would you need if the slope was going to be equal to speed in meters per second?
11) What axes would you need if the slope was going to be the *rate of flow* of water in liters per minute?
12) What axes would you have if the slope was equivalent to an "Exchange Rate"?
13) Describe an easy method for drawing or sketching a graph using a table of 3 values.
14) What difference is there in this method if it's not a straight line?
15) Describe the "$x = 0$, $y = 0$" method for drawing a straight line graph. Why doesn't this method work for anything except straight lines?
16) Explain what "$y = mx + b$" means, including the significance of m and b.
17) Describe a method for plotting a graph using "$y = mx + b$."
18) Explain how you can find the x- or y-intercept of a graph from its equation.
19) Describe how to get an equation from a straight line graph.
20) Describe how you'd get the equation of a straight line if you knew its slope and a point on the line.
21) What are *parallel* lines? How can you tell if two straight line equations represent parallel lines?
22) How can you change the equation $y = mx + b$ to "move" its graph up by 3? What's the fancy math word that means "move"?
23) What are the 2 different types of curves that you should know the basic shape of?
24) What type of equation produces a parabola, or u-shaped graph?
25) What about an n-shaped graph?
26) What type of equation produces a corner-to-corner graph with a flat bit in the middle?
27) Do 2 examples of each of the following graphs, giving the equations, and sketching the graphs: a) u-shaped, b) n-shaped, c) corner-to-corner with a flat bit in the middle.
28) Describe 2 common methods for getting answers from a graph or graphs.
29) What does the slope of a travel graph mean?

Substituting Values into Formulas

This topic is a lot easier than you think.

$$F = \frac{9}{5}C + 32$$ ← This formula will do as an example.

Generally speaking, algebra is a pretty grim subject, but you should realize that some parts of it are VERY easy, and this is definitely the easiest part of all, so whatever you do, don't mess up on the easy stuff.

Method:

If you don't follow this STRICT METHOD, you'll just keep getting them wrong — it's as simple as that.

1) Write out the formula. e.g. $F = \frac{9}{5}C + 32$

2) Write it again, directly underneath, but substituting numbers for letters on the RHS. (Right-Hand Side) if C = 15, then $F = (\frac{9}{5} \times 15) + 32$

3) Work it out IN STAGES. Use PEMDAS to work things out IN THE RIGHT ORDER. WRITE DOWN values for each part as you go along.
$F = 27 + 32$
$= 59$
$F = 59$

PEMDAS

Parentheses, Exponents, Multiplication, Division, Addition, Subtraction

PEMDAS tells you the ORDER in which these operations should be done:
Work out parentheses first, then exponents (things like squaring), then multiply/divide groups of numbers, before adding or subtracting them.
This set of rules works really well for simple cases, so remember the word: PEMDAS.

Example:

A mysterious quantity, T, is given by: $T = (P - 7)^2 + \frac{4R}{Q}$
Find the value of T when P = 4, Q = -2, and R = 3

ANSWER:

1) Write down the formula: $T = (P - 7)^2 + \frac{4R}{Q}$
2) Put the numbers in: $T = (4 - 7)^2 + \frac{4 \times 3}{-2}$
3) Then work it out in stages: $= (-3)^2 + \frac{12}{-2}$
$= 9 + -6$
$= 9 - 6 = 3$

Note PEMDAS in operation:
Parentheses worked out first, then squared (exponents).
Multiplications and divisions done before finally adding and subtracting.

The Acid Test:

LEARN the 3 Steps of the Substitution Method and the full meaning of PEMDAS. Then turn over...

...and write it all down from memory.

1) Practice the above example until you can do it easily without help.
2) If $C = \frac{5}{9}(F - 32)$, find the value of C when F = 77.

Solving Equations the Easy Way

The "proper" way to solve equations is shown on P.46, but in practice the "proper" way can be pretty difficult, so there's a lot to be said for the much easier methods shown below.

The drawback with these is that you can't always use them on very complicated equations. In most Exam questions though, they do just fine.

1) The "Common Sense" Approach

The trick here is to realize that the unknown quantity "x" is, after all, just a number, and the "equation" is just a cryptic clue to help you find it.

Example: *"Solve this equation: $3x + 4 = 46$"*
(i.e. find what number x is)

Answer: *This is what you should say to yourself:*

"Something + 4 = 46" hmm, so that "something" must be 42.

So that means $3x = 42$, which means "3 times something = 42."

So it must be $42 \div 3$ which is 14 so "$x = 14$."

In other words, don't think of it as algebra, but as "Find the mystery number."

2) The Trial and Error Method

This is a perfectly good method, and although it won't work every time, it usually does, especially if the answer is a whole number.

The *big secret of trial and error* methods is to find TWO OPPOSITE CASES, then keep trying values IN BETWEEN them.

In other words, find a number that makes the RHS bigger, and then one that makes the LHS bigger, and then try values in between them.

Example: *"Solve for x: $3x + 5 = 21 - 5x$"* (i.e. find the number x)

Answer:

Try $x = 1$: $3 + 5 = 21 - 5$, $8 = 16$ — no good, RHS too big

Try $x = 3$: $9 + 5 = 21 - 15$, $14 = 6$ — no good, now LHS too big

SO TRY IN BETWEEN: $x = 2$: $6 + 5 = 21 - 10$, $11 = 11$, YES, so $x = 2$.

"Trial and error" can be pretty useful for multiple choice questions too — e.g. if you're given five possible equations to match some numbers, just stick the numbers into each equation in turn and see which one works.

The Acid Test: LEARN these two methods until you can turn the page and write them down with an example for each.

Solve these equations using one of the methods above:

1) $4x - 12 = 20$ 2) $3x + 5 = 5x - 9$ 3) $2x - 7 = 20 - 7x$

Basic Algebra

1) Terms

Before you can do anything else, you MUST understand what a TERM is:

1) A TERM IS A COLLECTION OF NUMBERS, LETTERS, AND PARENTHESES, ALL MULTIPLIED/DIVIDED TOGETHER. (These are all terms.)

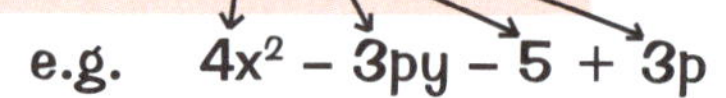

2) TERMS are SEPARATED BY + AND – SIGNS, e.g. $4x^2 - 3py - 5 + 3p$

3) TERMS always have a + or – ATTACHED TO THE FRONT OF THEM.

4) e.g. $4xy + 5x^2 - 2y + 6y^2 + 4$

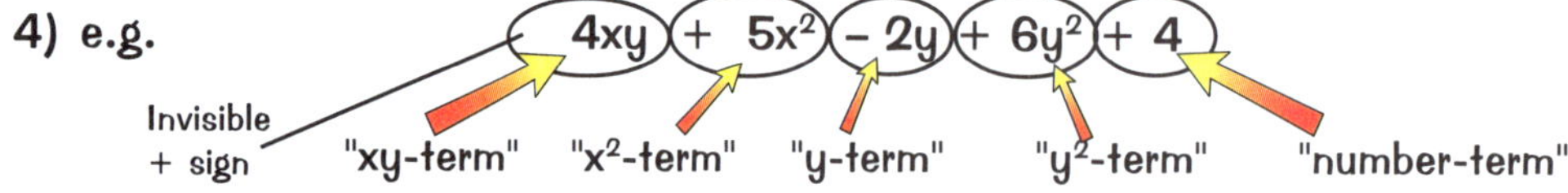

2) Simplifying

"Collecting Like Terms"

1) Put bubbles around each term — be sure you *capture the +/– sign* IN FRONT *of each*.
2) Then you can *move the bubbles into the best order* so that LIKE TERMS *are together*.
3) "LIKE TERMS" have exactly the same combination of letters, e.g. "x-terms" or "xy-terms."
4) Combine LIKE TERMS.

EXAMPLE: *"Simplify $2x - 4 + 5x + 6$"*

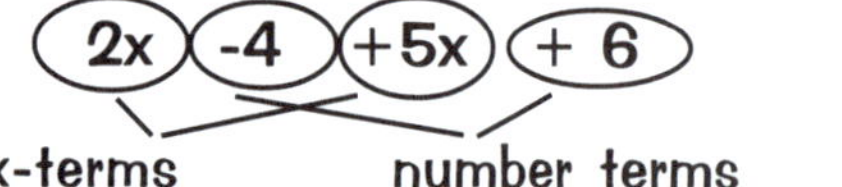

$2x \; -4 \; +5x \; +6 \;=\; +2x \; +5x \;\; -4 \; +6$

$= 7x \;\; +2 \;=\; 7x + 2$

3) Canceling Algebraic Fractions

This is exactly the same as canceling ordinary fractions.

1) Look for any bits that look the same (common factors) that are on *both the top and the bottom*.
2) Cancel them.

EXAMPLES: 1) Simplify: $\frac{6x(x+4)(x-1)}{3x(x-1)}$ Answer: $\frac{2\cancel{6x}(x+4)\cancel{(x-1)}}{\cancel{3x}\cancel{(x-1)}} = 2(x+4)$

2) Simplify: $\frac{2x^2y + 4xy^2 - 8xy}{2xy}$ Answer: $\frac{2x^2y + 4xy^2 - 8xy}{2xy} = \frac{\cancel{2xy}(x+2y-4)}{\cancel{2xy}} = (x+2y-4)$

You can only cancel if things are multiplied together. You need to factor the top line first — see next page.

The top line just has things multiplied together now — so you can cancel this.

4) Multiplying out Parentheses (Distribution)

1) The thing OUTSIDE the parentheses multiplies each separate term INSIDE the parentheses.
2) When letters are multiplied together, they are just written next to each other, e.g. pq.
3) Remember, $R \times R = R^2$, and TY^2 means $T \times Y \times Y$, while $(TY)^2$ means $T \times T \times Y \times Y$.
4) A minus outside the parentheses REVERSES ALL THE SIGNS when you multiply.

EXAMPLES:

1) $3(2x + 5) = 6x + 15$

2) $4p(3r - 2t) = 12pr - 8pt$

3) $-4(3p^2 - 7q^3) = -12p^2 + 28q^3$ (Note both signs have been reversed.)

Basic Algebra

5) Expanding and Simplifying

a) With Double Parentheses

— you get 4 terms after multiplying them out, and usually 2 of them combine to leave 3 terms, like this:

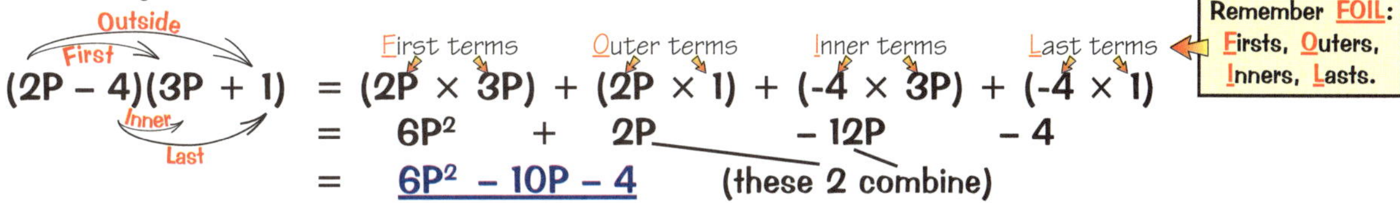

$$(2P - 4)(3P + 1) = (2P \times 3P) + (2P \times 1) + (-4 \times 3P) + (-4 \times 1)$$
$$= 6P^2 + 2P - 12P - 4$$
$$= 6P^2 - 10P - 4$$

(these 2 combine)

b) Squared Parentheses

e.g. $(3d + 5)^2$ ALWAYS write these out as two pairs of parentheses: $(3d + 5)(3d + 5)$ and work them out CAREFULLY like this:

$$(3d + 5)(3d + 5) = 9d^2 + 15d + 15d + 25 = 9d^2 + 30d + 25$$

YOU SHOULD ALWAYS GET FOUR TERMS from squared parentheses, and inevitably *two of these* will combine to leave THREE TERMS IN THE END, as shown above.

(The usual WRONG ANSWER, by the way, is $(3d + 5)^2 = 9d^2 + 25$ — eeek!)

6) Factoring — putting parentheses in

This is the *exact opposite* of multiplying out parentheses. Here's the method to follow:

1) Take out the biggest NUMBER that goes into all the terms.
2) Take each letter in turn and take out the highest power (e.g. x, x^2, etc.) that will go into EVERY term.
3) Open the parentheses and fill in all the parts needed to reproduce each term.

EXAMPLE: Factor $15x^4y + 20x^2y^3z - 35x^3yz^2$

Answer: $5x^2y(3x^2 + 4y^2z - 7xz^2)$

- Biggest number that'll divide into 15, 20, and 35.
- Highest powers of x and y that will go into all three terms.
- z wasn't in ALL terms so it can't come out as a common factor.

REMEMBER:

1) The parts taken out and put at the front are the COMMON FACTORS.
2) The parts inside the parentheses are what's needed to get back to the original terms (if you were to multiply the parentheses out again).

The Acid Test:

LEARN the important details for each of the 6 sections on these 2 pages, then turn over and write it all down.

Then apply the methods to these:

1) Simplify: a) $5x + 3y - 4 - 2y - x$ b) $4k + 3y^2 - 6k + y^2 + 2$ c) $\frac{4x^2y^3(z-1)}{2xy^2(z-1)^2}$

2) Expand: a) $2pq(3p - 4q^2)$ b) $(2g + 5)(4g - 2)$ c) $(4 - 3h)^2$

3) Factor: a) $14x^2y^3 + 21xy^2 - 35x^3y^4$ b) $12h^2j^3 + 6h^4j^2k - 36h^3jk$

Solving Equations

Solving Equations means finding the value of, say, x from something like: $3x + 5 = 4 - 5x$.
Now, not a lot of people know this, but exactly the same method applies to both solving equations and rearranging formulas as illustrated on these two pages.

1) EXACTLY THE SAME METHOD APPLIES TO BOTH FORMULAS AND EQUATIONS.
2) THE SAME SEQUENCE OF STEPS APPLIES EVERY TIME.

To illustrate the sequence of steps, we'll use this equation: $\sqrt{2 - \frac{x+4}{2x+5}} = 3$

The Six Steps Applied to Equations

1) Get rid of any square root signs by squaring both sides:

$$2 - \frac{x+4}{2x+5} = 9$$

2) Get everything off the bottom by cross-multiplying up to EVERY OTHER TERM:

$$2 - \frac{x+4}{2x+5} = 9 \Rightarrow 2(2x+5) - (x+4) = 9(2x+5)$$

3) Multiply out any parentheses:

$$4x + 10 - x - 4 = 18x + 45$$

4) Collect all subject terms on one side of the "=" and all nonsubject terms on the other side, remembering to reverse the +/– sign of any term that crosses the "=":

+18x moves across the "=" and becomes -18x
+10 moves across the "=" and becomes -10
-4 moves across the "=" and becomes +4

+/– +/– +/–

$$4x - x - 18x = 45 - 10 + 4$$

Subject terms have the unknown letter in... ...nonsubject terms are just numbers.

5) Combine like terms on each side of the equation, and reduce it to the form "Ax = B," where A and B are just numbers (or bunches of letters in the case of formulas):

$$-15x = 39$$

("Ax = B": A = -15, B = 39, x is the subject)

6) Finally slide the A underneath the B to give "$x = \frac{B}{A}$," divide (or cancel), and that's your answer:

$$x = -\frac{39}{15} = -2.6$$

So $x = -2.6$

The Acid Test: LEARN the 6 STEPS for solving equations and rearranging formulas. Turn over and write them down.

Solve the following equations: a) $5(x + 2) = 8 + 4(5 - x)$ b) $\frac{4}{x+3} = \frac{6}{4-x}$

Rearranging Formulas

Rearranging Formulas means making one letter (variable) the subject,
e.g. getting "y=" from something like $2x + z = 3(y + 2p)$.
Generally speaking, "solving equations" is easier, but don't forget:

1) EXACTLY THE SAME METHOD APPLIES TO BOTH FORMULAS AND EQUATIONS.
2) THE SAME SEQUENCE OF STEPS APPLIES EVERY TIME.

We'll illustrate this by making "y" the subject of this formula: $M = \sqrt{2K - \frac{K^2}{2y+1}}$

The Six Steps Applied to Formulas

1) Get rid of any square root signs by squaring both sides:

$$M^2 = 2K - \frac{K^2}{2y+1}$$

2) Get everything off the bottom by cross-multiplying up to EVERY OTHER TERM:

$$M^2 = 2K - \frac{K^2}{2y+1} \Rightarrow M^2(2y+1) = 2K(2y+1) - K^2$$

3) Multiply out any parentheses:

$$2yM^2 + M^2 = 4Ky + 2K - K^2$$

4) Collect all subject terms on one side of the "=" and all nonsubject terms on the other side, remembering to reverse the +/– sign of any term that crosses the "=":

$+4Ky$ moves across the "=" and becomes $-4Ky$
$+M^2$ moves across the "=" and becomes $-M^2$

$$2yM^2 - 4Ky = -M^2 + 2K - K^2$$

Subject terms contain the letter you're interested in... ...nonsubject terms don't.

5) Combine like terms on each side of the equation and reduce it to the form "Ay = B," where A and B are just bunches of letters which DON'T include the subject (y). Note that the left-hand side has to be FACTORED:

$$(2M^2 - 4K)y = 2K - K^2 - M^2$$

("Ay = B" i.e. $A = (2M^2 - 4K)$, $B = 2K - K^2 - M^2$, y is the subject)

6) Finally slide the A underneath the B to give "$y = \frac{B}{A}$," (cancel if possible) and that's your answer:

So $$y = \frac{2K - K^2 - M^2}{(2M^2 - 4K)}$$

The Acid Test:

STILL KNOW the 6 STEPS? Make sure you can do the example above without looking at the page.

1) Rearrange "$F = \frac{9}{5}C + 32$" from "F=" to "C=" and then back the other way.

2) Make p the subject of these: a) $\frac{p}{p+y} = 4$ b) $\frac{1}{p} = \frac{1}{q} + \frac{1}{r}$

Concept — Algebra and Functions
Use variables and appropriate operations to write an expression, an equation, an inequality, or a system of equations that represents a verbal description.
Concept — Mathematical Reasoning Analyze problems by identifying relationships...
Algebra 1 Students solve multistep problems, including word problems, involving linear equations and linear inequalities in one variable and provide justification for each step.

Word Problems

This topic is the key to life, the Universe, and everything.

Well alright, maybe I've overstated it a little, but it's pretty darned important, OK.
A lot of the time, math problems are cunningly disguised as ordinary sentences.
They don't give you an equation — you have to FIGURE IT OUT FOR YOURSELF. Here's how.

Method:

IGNORE this method and you'll probably just end up STARING at a question until the END OF TIME.

1) List all the quantities you're given or that you have to find out.
2) Write down a symbol for each quantity.
   (If you're not told them in the question, you'll have to make them up.)
3) Write down the values of any symbols you know.
4) If you can, draw a diagram.
5) Now look at the symbols and your diagram (if you drew one), and write down all the relationships between them (but only using math symbols).
   (There — now you've got your equation.)
6) Now the easy part — SOLVE the equation.

You can also use this method to get systems of equations (see P.50). You just do it exactly the same way, except at step 5 you'll find you need to write down more than one equation.

A shop sells a stereo system for \$21 more than twice the amount it costs the shop. If it sells the stereo system for \$105, then what did the shop pay for it?

ANSWER:

1) List the quantities: the cost to the shop, and the cost to the customer
2) Write down the symbols: You're not told any, so you'll have to make them up:
   I think I'll choose S = cost to shop and C = cost to customer.
3) Write down the values you know: $C = 105$
4) Draw a diagram:

C = 105 | 21 | S | S

You might've thought you couldn't draw a diagram, but just take a look at this — and tell me if it doesn't make it easier.

5) The relationship: $C = 21 + 2S$ (easy if you drew the diagram)
6) Solve it:
   First stick in $C = 105$: $105 = 2S + 21$
   Rearrange: $2S = 105 - 21 = 84$
   Divide by 2: $S = 42$

Check that your answer sounds reasonable — this is just under half the selling price, which is about what you would have expected.

So the shop paid \$42 for it.

Word Problems

Another Example:

Sarah is 20 miles from a youth hostel. She walks directly toward it at a steady pace of 4 miles per hour. Write down an equation giving Sarah's distance in miles from the hostel, y, after x hours.

ANSWER:

1) List the quantities: the starting distance, the distance at a given time, the time, and the speed.

2) Write down the symbols:

OK, we're given x = time and y = distance at that time,
and I think I'll use d = starting distance and s = speed.

3) Write down the values you know: $s = 4$ $d = 20$

See P.16 for more about speed.

4) Draw a diagram:

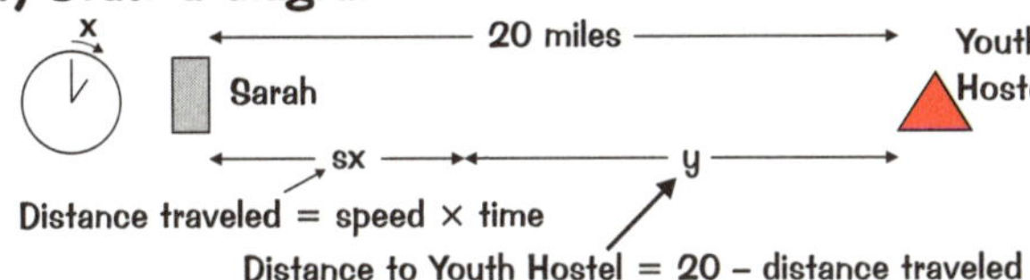

5) The relationship: $y = d - sx$

6) Now just stick in the figures you know:

$y = 20 - 4x$

Yet Another Example:

A rectangular garden is 17 feet wide, and has an area of 391 square feet. If the garden is p feet longer than it is wide, write down and solve an equation for p.

ANSWER:

1) List the quantities: area, length, and width.

2) Write down the symbols:

We're given p = the difference between the length and the width,
while A = area, x = width, and y = length seem obvious choices.

3) Write down the values you know: $A = 391$ $x = 17$

When there's more than one equation, it's called a "system of equations" (see P.50)

4) Draw a diagram:

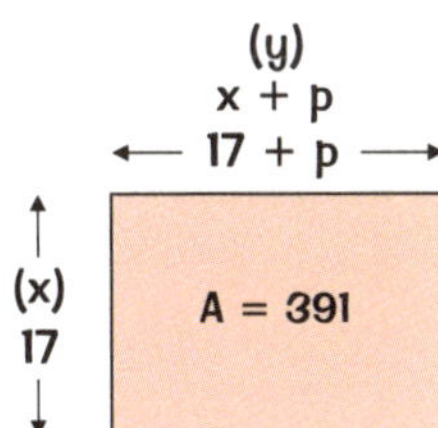

5) The relationships: $A = xy$ where $y = x + p$

...so replace the "y" in this first equation with "x + p":
(just stick one equation "inside the other") $A = x(x + p)$

6) Now stick in A = 391 and x = 17 to give:

$$17(17 + p) = 391$$
$$289 + 17p = 391$$
$$17p = 102$$
$$p = 6$$

So the garden is 6 feet longer than it is wide — which seems reasonable. You'd probably worry if you got an answer like 40,000 or 0.005, but 6 feet sounds okay.

The Acid Test:

Turn over and make sure you can write down the essential 6-point method.

Then try your luck with these:

1) Esmerelda spends $\frac{2}{5}$ of her monthly income on her rent. If her income is $350, then how much is her rent?
2) Luke bought some socks for $3 a pair and some shirts for $12 each. If he spent $96 and bought 2 more pairs of socks than shirts, how many shirts did he buy?
3) A rectangular field is twice as long as it is wide and has a perimeter of 100 yards. Write down a system of equations that could be used to find the length "l" and width "w" of the field.

Students solve a system of of two linear equations in two variables algebraically and are able to interpret the answer graphically. Students are able to solve a system of two linear inequalities in two variables and to sketch the solution sets.

Systems of Linear Equations

These are OK as long as you learn these SIX STEPS in every meticulous detail.

The Six Steps:

You need these two equations for the example:
$2x = 6 - 4y$ and $-3 - 3y = 4x$

1) REARRANGE BOTH EQUATIONS into the form: $Ax + By = C$ where A, B, and C are numbers (which can be negative). Also LABEL the two equations ① and ②.

$2x + 4y = 6$ ①
$-4x - 3y = 3$ ②

2) You need to MATCH UP THE NUMBERS IN FRONT (the "coefficients") of either the x's or y's IN BOTH EQUATIONS. To do this you may need to MULTIPLY one or both equations by a suitable number. You should then RELABEL them: ③ and ④.

2 × ① : $4x + 8y = 12$ ③
$-4x - 3y = 3$ ④

(This gives you +4x in equation ③ to match the -4x in equation ②, now called ④.)

3) ADD OR SUBTRACT the two equations... ...to eliminate the terms with the same coefficient. If the coefficients are the SAME (both positive or both negative) then SUBTRACT. If the coefficients are OPPOSITE (one positive and one negative) then ADD.

③ + ④ $(+4x + -4x) + (8y + -3y) = (12 + 3)$
$0x + 5y = 15$

(In this case, you have +4x and -4x so you ADD.)

4) SOLVE the resulting equation to find whichever letter is left in it.

$5y = 15 \Rightarrow y = 3$

5) SUBSTITUTE this back into equation ①, and solve it to find the other quantity.

Sub in ①: $2x + (4 \times 3) = 6 \Rightarrow 2x + 12 = 6 \Rightarrow 2x = -6 \Rightarrow x = -3$

6) Then SUBSTITUTE BOTH THESE VALUES INTO EQUATION ② to make sure it gives the right answer. If it doesn't, then you've done something wrong and you'll have to do it all again.

Sub x and y in ②: $(-4 \times -3) - (3 \times 3) = 12 - 9 = 3$ which is right, so it's worked.

So the solutions are: $x = -3$, $y = 3$

The Acid Test: LEARN the 6 Steps for solving Systems of Linear Equations.

Remember, you only know them when you can write them all out from memory, so turn over the page and try it. Then apply the 6 steps to find F and G given that
$2F - 10 = 4G$ and $3G = 4F - 15$.

Graphing Systems of Linear Equations

On the opposite page is the tricky algebra method for solving systems of equations.
On this page is the nice easy graph method for solving them.
You could be asked to do either method in the Exam, so make sure you learn them both.

Solving Systems of Equations Using Graphs

This is a very easy way to find the x- and y-solutions of a system of two equations.
Here's the simple rule:

THE SOLUTION OF A SYSTEM OF TWO EQUATIONS IS SIMPLY THE X- AND Y-VALUES WHERE THEIR GRAPHS CROSS

Three-Step Method:

1) Do a *"TABLE OF 3 VALUES"* for both equations.
2) Draw the *TWO GRAPHS*.
3) Find the x- and y-values *WHERE THEY CROSS*.

Piece o' cake.

Example: *"Draw the graphs for "$y = 2x + 3$" and "$y = 6 - 4x$" and then use your graphs to solve the system of equations."*

1) TABLE OF 3 VALUES (see P.34) for both equations:

x	0	2	-2
y	3	7	-1

x	0	2	3
y	6	-2	-6

2) DRAW THE GRAPHS:

3) WHERE THEY CROSS, $x = \frac{1}{2}$, $y = 4$.

And that's the answer...

$x = \frac{1}{2}$ and $y = 4$

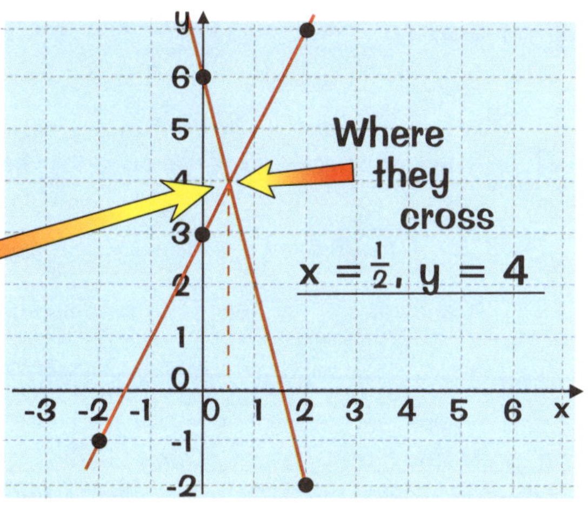

The Acid Test:

LEARN the Simple Rule and the 3-step method for solving systems of equations using GRAPHS...

...now cover the page and write down what you've learned from memory.

1) Use graphs to find the solutions to these pairs of equations:

a) $y = 4x - 4$ and $y = 6 - x$ b) $y = 2x$ and $y = 6 - 2x$

Inequalities

$>$ means "Greater than" $\geq$ means "Greater than or equal to"
$<$ means "Less than" $\leq$ means "Less than or equal to"

Inequalities like to be put on Number Lines:

It's true. Putting an inequality on a number line is like taking your dog to its favorite park.

EXAMPLES:

$x > 5$

-10 -9 -8 -7 -6 -5 -4 -3 -2 -1 0 1 2 3 4 5 6 7 8 9 10

$x \leq -2$

-10 -9 -8 -7 -6 -5 -4 -3 -2 -1 0 1 2 3 4 5 6 7 8 9 10

$-8 < x \leq 2$

-10 -9 -8 -7 -6 -5 -4 -3 -2 -1 0 1 2 3 4 5 6 7 8 9 10

● includes the value
○ doesn't include the value

Algebra with Inequalities — generally a little tricky.

The thing to remember here is that inequalities are just like regular equations... in the sense that all the normal rules of algebra apply... ...but with ONE BIG EXCEPTION:

Whenever you MULTIPLY OR DIVIDE BY A NEGATIVE NUMBER, you must FLIP THE INEQUALITY SIGN.

Example: "Solve $5x < 6x + 2$"

ANSWER: First move the 6x over the "$<$": $5x - 6x < 2$

combining the x-terms gives: $-x < 2$

To get rid of the "–" in front of x you need to divide both sides by -1 — but remember that means the "$<$" has to be flipped as well, which gives:

$x > -2$ i.e. "x is greater than -2" is the answer.

(The < has flipped around into a >, because we divided by a negative number.)

This answer, $x > -2$, can then be displayed on a number line like this:

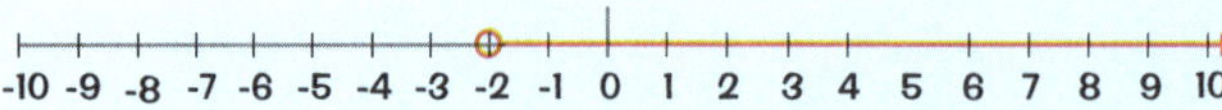

The main thing you should realize is that most of the time you just treat the "<" or ">" as though it was an "=" and do all the usual algebra that you would for a regular equation. The "Big Exception" doesn't actually come up very often at all.

The Acid Test:

LEARN: The 4 Inequality Signs, the similarity with EQUATIONS, and the One Big Exception.

Now turn over and write down what you've learned.

1) Solve this inequality: $4x + 3 \leq 6x + 7$.
2) Find all the integer values of x which satisfy both $2x + 9 \geq 1$ and $4x < 6 + x$.

Students solve multistep problems, including word problems, involving linear equations and linear inequalities in one variable and provide justification for each step. Students solve a system of of two linear equations in two variables algebraically and are able to interpret the answer graphically. Students are able to solve a system of two linear inequalities in two variables and to sketch the solution sets.

Graphing Inequalities

This page looks a lot worse than it actually is. The method seems really long-winded at first... But with a little practice, you'll realize there's really nothing to it at all.

Method:

1) **CONVERT each INEQUALITY to an EQUATION**

by simply putting an "=" in place of the "<" or whatever.

2) **DRAW THE LINE FOR EACH EQUATION**

3) **CHOOSE A NICE POINT THAT ISN'T ON THE LINE**

The ORIGIN is a nice one — you'll see why in a minute. But remember you need a point NOT on the line, so you CAN'T ALWAYS use the origin.

4) **SEE IF IT'S ON THE CORRECT SIDE OF EACH LINE**

Stick the x- and y-values of your point into each inequality. (That's why the origin's nice — x and y are 0, which is much easier.) If the inequality is TRUE, the point's on the correct side of the line. If it's FALSE, the point's on the wrong side.

5) **THIS GIVES YOU THE REGION — SHADE IT**

Example:

"Shade the region represented by: $x + y \leq 5$ and $y > x + 2$"

ANSWER:

1) CONVERT EACH INEQUALITY TO AN *EQUATION*:

$x + y \leq 5$ becomes $x + y = 5$.

$y > x + 2$ becomes $y = x + 2$.

2) DRAW THE LINE for each equation

Either draw a table of values (P.34) or use "$y = mx + b$" (P.35).

A solid line means "includes the value." (i.e. $\leq$ or $\geq$)

A dotted line means "doesn't include the value." (i.e. < or >)

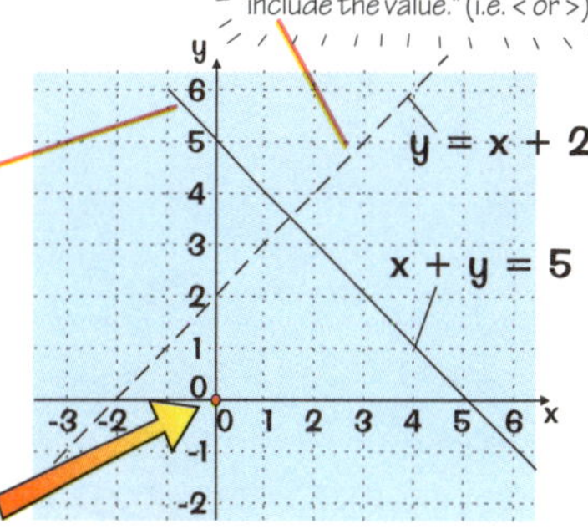

3)  CHOOSE A POINT,

Well if I must...I choose point (0, 0).
(It's underneath the line $x + y = 5$ and underneath $y = x + 2$. See...)

4)  SEE IF IT'S ON THE CORRECT SIDE OF THE LINES

Stick $x = 0$ and $y = 0$ into each inequality and see if they're true:

$x + y \leq 5$ becomes $0 \leq 5$ which is TRUE
$y > x + 2$ becomes $0 > 2$ which is FALSE, so we want the other side.

So the region we want is UNDERNEATH $x + y = 5$
but also ABOVE $y = x + 2$.

5)  SHADE THE FINAL REGION, and there you have it.

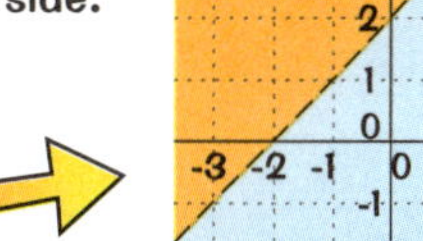

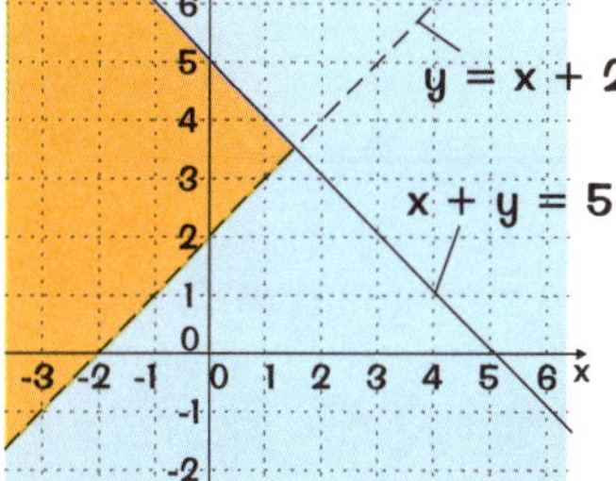

The Acid Test:

LEARN the Five Steps for graphing inequalities, then turn over and write them down.

1) Show on a graph the region which satisfies the following three conditions:

$x + y < 6, \quad y > 0.5, \quad y < 2x - 2$

Concept — Number Sense
Understand the meaning of the absolute value of a number; interpret the absolute value as the distance of the number from zero on a number line; and determine the absolute value of real numbers.

Algebra 1 Students solve equations and inequalities involving absolute values.

Questions with Absolute Values

These can be a bit of a pain unless you take your time, so just make sure at each stage that what you're doing makes sense. (For more about what an absolute value is, see P.11.)

You get a Two-Part Answer to Absolute Value Questions

Answers to an absolute value question always have two parts. Basically, you don't know whether the bit in the absolute value sign is positive or negative — so you have to allow for both possibilities.

EXAMPLES:

① $|x| = 3$

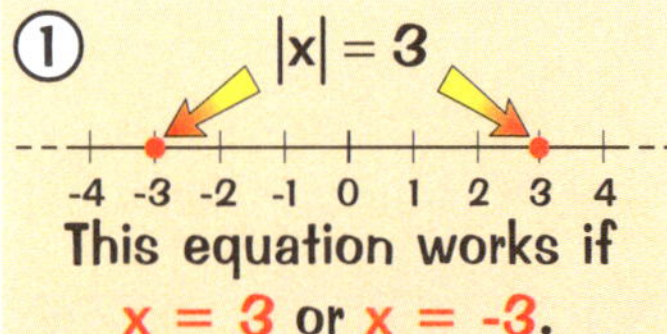

This equation works if $x = 3$ or $x = -3$.

② $|m| > 2$

-5 -4 -3 -2 -1 0 1 2 3 4 5

This inequality is true if $m < -2$ or $m > 2$.

③ $|p| \le 4$

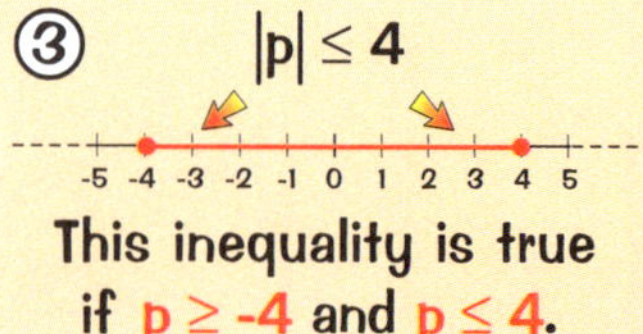

This inequality is true if $p \ge -4$ and $p \le 4$.

Write Down Absolute Value Questions... Twice

If you get an equation or inequality with the absolute value sign around, say, the letter x...

1) Write down TWO VERSIONS of the equation/inequality:
   (i) in the first, WRITE x INSTEAD OF $|x|$ (i.e. assume x takes a POSITIVE value).
   (ii) in the second, WRITE -x INSTEAD OF $|x|$ (i.e. assume x takes a NEGATIVE value).
   (NOTE: With things like $|a - b|$, you have to make the WHOLE EXPRESSION negative — see example c.)
2) SOLVE each of these new equations/inequalities SEPARATELY.

EXAMPLES: Solve a) $3|x| + 7 = 16$ b) $6 + 2|m| \le 10$ and c) $|3y - 4| > 8$

a) The two versions:

(i) $3x + 7 = 16$
(ii) $3(-x) + 7 = 16 \longrightarrow -3x + 7 = 16$

Solving (i):
$3x + 7 = 16$
$3x = 9$
$x = 3$

Solving (ii):
$-3x + 7 = 16$
$-3x = 9$
$x = -3$

b) The two versions:

(i) $6 + 2m \le 10$
(ii) $6 + 2(-m) \le 10 \longrightarrow 6 - 2m \le 10$

Solving (i):
$6 + 2m \le 10$
$2m \le 4$
$m \le 2$

Solving (ii):
$6 - 2m \le 10$
$6 \le 10 + 2m$
$-4 \le 2m$
$-2 \le m$
i.e. $m \ge -2$

So m is greater than or equal to -2, but less than or equal to 2. If your answer has to be an integer (i.e. a whole number) then you could write this solution as {-2, -1, 0, 1, 2}.

c) The two versions:

(i) $3y - 4 > 8$
(ii) $-(3y - 4) > 8 \longrightarrow -3y + 4 > 8$

All of "3y – 4" is in the absolute value sign, so you have to make it all negative.

Solving (i):
$3y - 4 > 8$
$3y > 12$
$y > 4$

Solving (ii):
$-3y + 4 > 8$
$-3y > 4$
$y < -\frac{4}{3}$

You're dividing by -3, so don't forget to flip the inequality. (See P.52)

The Acid Test:

TRIAL & ERROR is a good way to solve these too (see P.43).

1) Solve: a) $3|x| - 4 = 8$ b) $5 - |q| < 4$ c) $2 + 4|p| < 10$.
2) Solve: a) $|k - 7| = 2$ b) $|n - 2| < 3$ c) $|m + 2| > 1$.

Review Summary for Section Four

It has to be said, Section Four is as grizzly as a bucket of grizzly stuff. But there are some pretty useful things in there. It'll come in handy wherever you are... particularly in your Exam. Remember the basic principle of learning — REPETITION IS GOOD. So keep doing these questions. Practice them 'til you can answer them ALL without referring to the section. Then go wrestle with a real grizzly. Or watch some TV.

Keep learning these basic facts until you know them

1) Describe how to substitute values into formulas.
2) What does PEMDAS have to do with putting numbers into formulas?
3) Describe 2 easier alternative methods for tackling simple equations.
4) Demonstrate your prowess at these methods by doing an example of each.
5) In algebra, what is a *term*?
6) Explain how to simplify algebraic expressions (collecting like terms).
7) Describe how to cancel algebraic fractions.
8) Explain how to multiply out parentheses (distribution).
9) Explain how to factor algebraic expressions.
10) What does solving equations have in common with rearranging formulas?
11) Give a 6-step method for solving equations and rearranging formulas.
12) Give 6 steps for solving a word problem.
13) Use the 6 steps to write the system of equations you'd need to solve this problem: *"Tony has a 7-foot long utility shelf in his garage. He bought a $12\frac{1}{2}$-ft long piece of wood to extend it. He adds a length of wood to each end of the shelf so that it runs all along one wall of his garage. The lengths of wood he adds are x inches and (x + 27) inches long and he has 3 ft 9 in of wood left when he's done. Find x and the length of the garage, y."*
14) What are systems of linear equations? Give an example.
15) Describe a method for solving systems of linear equations using algebra.
16) Describe a method for solving systems of linear equations using graphs.
17) What are the 4 inequality signs, and what do they mean?
18) How should you represent each of the 4 types of inequalities on a number line?
19) What are the rules of algebra for inequalities (including the big exception)?
20) Describe a method for graphing inequalities.
21) What's the extra complication with equations and inequalities that contain absolute values?
22) Explain how you find the solutions to an equation or inequality containing absolute values.

Represent probabilities as ratios, proportions, decimals between 0 and 1, and percentages between 0 and 100 and verify that the probabilities computed are reasonable; know that if P is the probability of an event, 1-P is the probability of an event not occurring.

Understand the difference between independent and dependent events.

Probability

This is nobody's favorite subject — for sure, I've never really spoken to anyone who's said they like it (not for long anyway).

Actually, it's not as bad as you might think, but YOU MUST LEARN THE BASIC FACTS, which are what you have on these 3 pages.

All Probabilities are between 0 and 1

A probability of ZERO means it will NEVER HAPPEN.

A probability of ONE means it DEFINITELY WILL.

You can't have a probability bigger than 1.

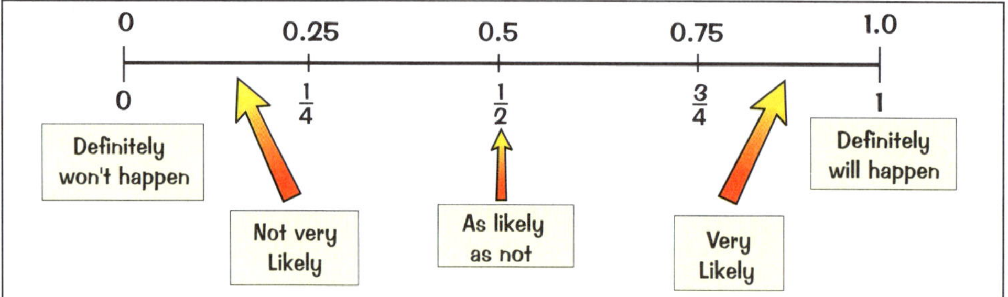

You should be able to put the probability of any event happening on this scale of 0 to 1.

Three Important Details

1) PROBABILITIES SHOULD BE GIVEN as either A FRACTION ($\frac{1}{4}$), A DECIMAL (0.25), or A PERCENT (25%)

2) THE NOTATION : "P(X) = $\frac{1}{2}$" SHOULD BE READ AS: "The probability of event X happening is $\frac{1}{2}$."

3) PROBABILITIES ALWAYS ADD UP TO 1. This is essential for finding the probability of *the other outcome.* e.g. If P(pass) = $\frac{1}{4}$, then P(fail) = $\frac{3}{4}$.

FOUR IMPORTANT EXTRAS — *for combined events (ready for page 58!):*

1) Watch out for "WITH REPLACEMENT" and "WITHOUT REPLACEMENT" and make sure you know what difference it makes.
(Either you put the thing back after the first go, before having your second go, or you don't — the 2nd tree diagram over the page illustrates what can happen.)

2) The COMBINED PROBABILITY of two events BOTH happening is ALWAYS LESS than the probability of either of them occurring alone.

3) EVENTS ARE INDEPENDENT if one event happening does not in any way affect the probability of the other one happening. Contrast this with *disjoint* below.

4) EVENTS ARE DISJOINT if one event happening means that the other one *can't* happen. Pretty much the opposite of *independent events* (see above).

The Acid Test:

LEARN the diagram and all the IMPORTANT POINTS on this page. Then turn over and write it all down.

1) Write 0.35 as a) a fraction, and b) a percent.

2) If P(picking a blue ball) is $\frac{1}{4}$, what is the value of P(not picking a blue ball)?

3) My drawer contains red, green, and blue socks. If the probability of picking out a red sock is $\frac{1}{4}$, and the probability of picking a green sock is $\frac{1}{2}$, what's the probability of picking a blue sock?

Calculating Probabilities

1) Equal Probabilities

When the different results all have the same chance of happening, then the probabilities will be EQUAL. These are the two cases which usually come up in the Exam:

1) TOSSING A COIN — Equal chance of getting a head or a tail ($\frac{1}{2}$).

2) THROWING A DICE — Equal chance of getting any of the numbers ($\frac{1}{6}$).

2) Unequal Probabilities You Can Work Out

These make for more interesting questions. (Which means you'll get them in the Exam.)

EXAMPLE: *"A bag contains 6 blue balls, 5 red balls, and 9 green balls. Find the probability of picking out a green ball."*

ANSWER: The chances of picking out the three colors are NOT EQUAL. The probability of picking a green is simply:

$$\frac{\text{NUMBER OF GREENS}}{\text{TOTAL NUMBER OF BALLS}} = \frac{9}{20}$$

3) Listing all Outcomes: 2 Coins, Dice, Spinners

If things are a little more complicated, you can work out probabilities by listing all the possible outcomes. This works if you've got 2 dice, a spinner and a coin, 2 spinners, etc.

EXAMPLE 1: What's the probability of getting a head and a tail when I toss 2 coins?

The possible outcomes from tossing two coins are:

1st coin \ 2nd coin	Head (H)	Tail (T)
Head (H)	HH	HT
Tail (T)	TH	TT

There are 4 possible outcomes, and 2 of these have 1 head and 1 tail.

So the probability of getting a head and a tail is:

$$\frac{\text{NUMBER OF OUTCOMES WITH 1 HEAD AND 1 TAIL}}{\text{TOTAL NUMBER OF POSSIBLE OUTCOMES}} = \frac{2}{4} = \frac{1}{2}$$

EXAMPLE 2: What's the probability of getting green and at least a 2 from these spinners?

List the possible outcomes in a table...

1st spinner \ 2nd spinner	Red	Green	Blue
1	1 + Red	1 + Green	1 + Blue
2	2 + Red	2 + Green	2 + Blue
3	3 + Red	3 + Green	3 + Blue

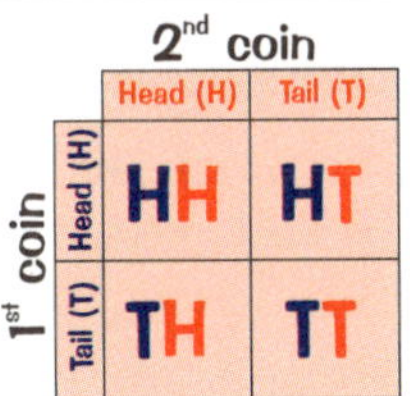

$$\text{Probability} = \frac{\text{NUMBER OF 'GOOD' OUTCOMES}}{\text{NUMBER OF POSSIBLE OUTCOMES}} = \frac{2}{9}$$

The Acid Test:

LEARN how to work out PROBABILITIES of events where you can LIST the possible outcomes.

1) What is the probability of throwing at least a three with a normal 6-sided dice?
2) What is the probability of scoring exactly 10 when I throw 2 normal 6-sided dice?

Probability — Tree Diagrams

General Tree Diagram

Tree diagrams are all pretty much the same, so it's a darned good idea to learn these basic details (which apply to ALL tree diagrams).

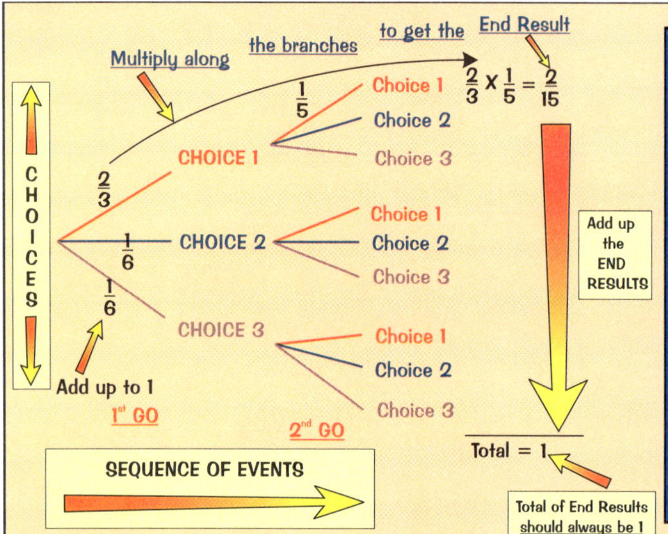

1) *Always MULTIPLY ALONG THE BRANCHES* (as shown) to get the END RESULTS.
2) *On any set of branches which all meet at a point*, the numbers must always ADD UP TO 1.
3) *Check that your diagram is correct* by making sure the End Results ADD UP TO ONE.
4) *To answer any question*, simply ADD UP THE RELEVANT END RESULTS (see below).

A Likely Tree Diagram Question

EXAMPLE: *"A box contains 3 red disks and 6 green disks. Two disks are taken without replacement. Draw a tree diagram and hence find the probability that both disks are the same color."*

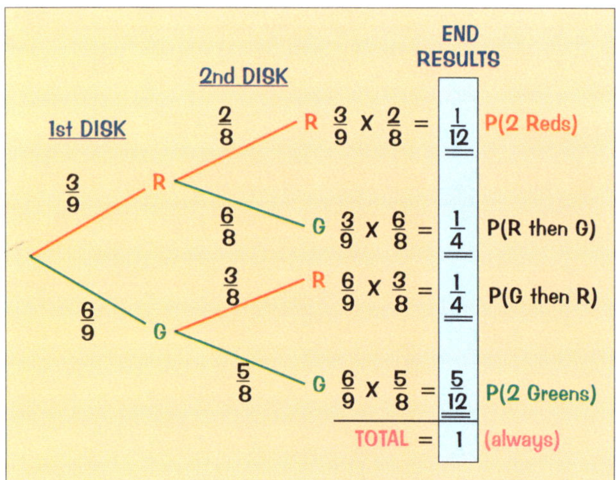

Once the tree diagram is drawn all you then need to do to answer the question is simply select the RELEVANT END RESULTS and then ADD THEM TOGETHER:

Probability of 2 Reds = $\frac{1}{12}$

Probability of 2 Greens = $\frac{5}{12}$

$$\frac{1}{12} + \frac{5}{12} = \frac{6}{12} = \frac{1}{2} = 0.5 \text{ (or } 50\%)$$

Remember, you can give probabilities using fractions, decimals, or percents.

The Acid Test:

LEARN the GENERAL TREE DIAGRAM, and the 4 points that go with it.

1) OK, let's see what you've learned shall we: *TURN OVER AND WRITE DOWN EVERYTHING YOU KNOW ABOUT TREE DIAGRAMS.*
2) A bag contains 6 red and 4 black tarantulas. A girl plucks out one tarantula at random, and then another. Draw a tree diagram to find the probability that she gets different colored spiders.

Mean, Median, and Mode

These three guys are all kinds of averages. And if you don't manage to learn these 3 basic definitions then you'll be passing up on some of the easiest points in any exam. It can't be that difficult can it?

1) MODE = MOST common

Mode = most (emphasize the 'o' in each when you say them)

2) MEDIAN = MIDDLE value

Median = mid (emphasize the m*d in each when you say them)

3) MEAN = TOTAL of items ÷ NUMBER of items

Mean is just the average, "but it's mean because you have to work it out"

The Golden Rule:

Mean, median, and mode should be easy points but even people who've gone to the incredible extent of learning them still manage to lose points in tests because they don't do this one vital step:

Always REARRANGE the data in INCREASING ORDER

(and check you have the same number of entries!)

Example: "Find the mean, median, and mode of these 14 numbers:"
2, 5, 3, 2, 6, -4, 0, 9, -3, 1, 6, 3, -2, 3

1) FIRST... rearrange them: -4, -3, -2, 0, 1, 2, 2, 3, 3, 3, 5, 6, 6, 9 (✓14)

2) MEAN $= \frac{\text{total}}{\text{number}} = \frac{-4-3-2+0+1+2+2+3+3+3+5+6+6+9}{14}$

$= 31 \div 14 = \underline{2.21}$

3) MEDIAN = the middle value (only when they're arranged in order of size, that is!).

When there are TWO MIDDLE NUMBERS, then the median is HALFWAY BETWEEN THE TWO MIDDLE NUMBERS

-4, -3, -2, 0, 1, 2, 2, 3, 3, 3, 5, 6, 6, 9
seven numbers this side ↑ seven numbers this side
Median = 2.5

4) MODE = most common value, which is simply 3. (Or you can say, "The modal value is 3.")

The Acid Test:

LEARN The Three Definitions and THE GOLDEN RULE...

...then cover this page and write them down from memory.

1) Apply all that you have learned to find the mean, median, and mode for this set of data:
1, 3, 14, -5, 6, -12, 18, 7, 23, 10, -5, -14, 0, 25, 8

Quartiles and Box-and-Whisker Plots

The median divides your data into two halves — an upper half and a lower half.
The quartiles do the same kind of thing, but they divide your data into 4 parts.

The Quartiles are Medians of Half the Data

1) Before you do anything else, follow the Golden Rule:
   PUT YOUR DATA IN INCREASING ORDER.

2) DIVIDE YOUR DATA INTO TWO HALVES — the LOWER HALF and the UPPER HALF.
   If you have an odd number of values, IGNORE THE MIDDLE VALUE (the median).

3) The LOWER QUARTILE is the MEDIAN OF THE LOWER HALF of values.
   The UPPER QUARTILE is the MEDIAN OF THE UPPER HALF of values.

Example: Here are 17 people's weights in lbs.
147, 149, 159, 160, 150, 140, 153, 141, 156, 155, 144, 144, 154, 157, 163, 148, 149.

1) The Golden Rule: Put the data in increasing order.

140	141	144	144	147	148	149	149	150	153	154	155	156	157	159	160	163

2) Divide the data into two halves.
   There's an odd number of values — so ignore the middle value (the median).

3) The lower quartile is midway between 144 and 147.

LOWER QUARTILE = 145.5

The upper quartile is midway between 156 and 157.

UPPER QUARTILE = 156.5

Box-and-Whisker Plots show the Median and Quartiles

Box-and-whisker plots are a neat graphical summary of your data.

TO CREATE A BOX-AND-WHISKER PLOT:

1) Draw the scale along the bottom.
2) Draw a box from the lower quartile to the upper quartile.
3) Draw a line down the box to show the median.
4) Draw "whiskers" out to the maximum and minimum.

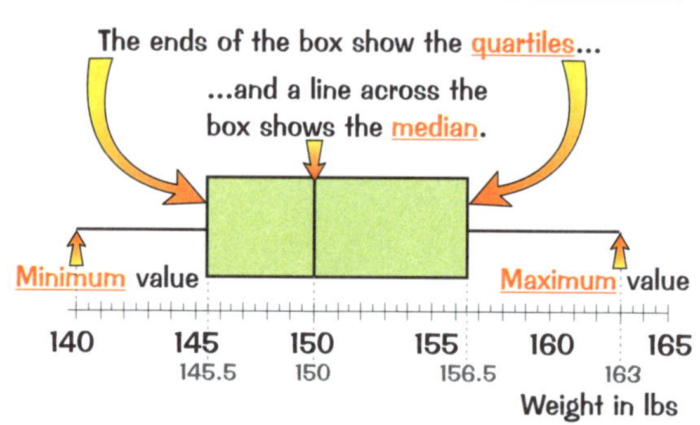

The Acid Test:

LEARN how to work out the UPPER AND LOWER QUARTILES, and how to draw a BOX-AND-WHISKER PLOT.

Find the median, and upper and lower quartiles of the following heights of 17 people (in inches).
70, 67, 61, 60, 65, 73, 70, 62, 63, 72, 62, 67, 60, 62, 70, 72, 71
Draw a box-and-whisker plot to summarize your results.

Interpreting Statistics

"There are lies, darned lies, and statistics." Kind of harsh maybe, but you know what he meant. Statistics can be used in misleading ways, so you need to question what's being claimed.

Two dead important questions to ask...

1) What's more suitable — an AVERAGE or a TOTAL?
2) Does this SAMPLE reflect the WHOLE POPULATION?

The Average can Give a Clearer Picture than the Total

EXAMPLE: Bryan Johnson and Rich Briars are basketball players. Bryan thinks he should be paid 3 times as much as Rich, since he's scored 3 times as many points, and so must be 3 times as good.

The owner of the team looks at the figures, then points out a few facts...

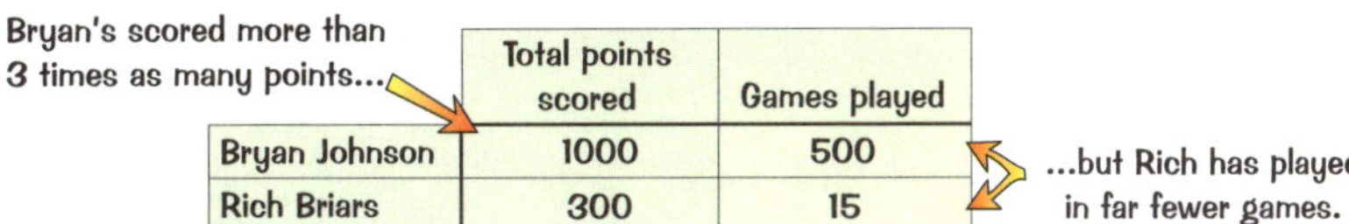

Bryan's scored more than 3 times as many points...

	Total points scored	Games played
Bryan Johnson	1000	500
Rich Briars	300	15

...but Rich has played in far fewer games.

The team-owner reckons Bryan's claim is misleading.

1) It's true that in total, Bryan's scored more than 3 times as many points as Rich.
2) But Rich has played in far fewer games — so it's better to look at the points scored per game.
3) This means working out the average number of points per game scored by each player.
4) On the average, Rich scores 10 times as many points per game as Bryan.

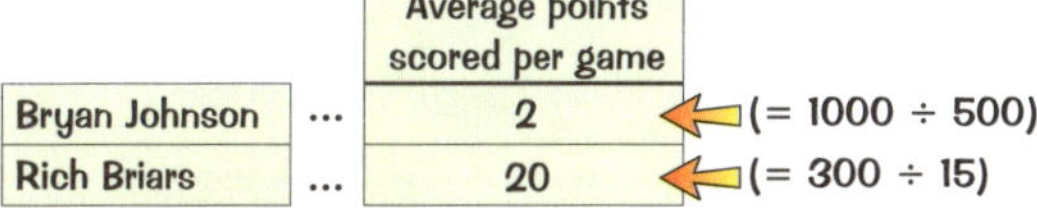

	Average points scored per game	
Bryan Johnson ...	2	(= 1000 ÷ 500)
Rich Briars ...	20	(= 300 ÷ 15)

Results from Samples can be Misleading

Surveys of "populations" (not necessarily people) are often necessary to find things out about them. But most times it's not possible to test the whole "population." In these cases a sample is tested, and the results are assumed to be true of the whole population. But there are potential problems.

In questions where samples have been used, you might have to "look for ways that the sample might not be a true reflection of the population as a whole."

Example:

In a telephone poll, 100 people were called at home and asked if they regularly watch daytime TV. A total of 80% said yes. Does this mean 80% of the population regularly watch daytime TV?

ANSWER: Probably not. There are several things wrong with this sampling technique:

1) First and worst: the sample is far too small. At least 1000 would be more like it.
2) What about people who don't have their own phone?
3) What time of day was it done? At 11 a.m. you'd probably get different results than at 10 p.m.
4) Which part or parts of the country would you telephone?

The Acid Test:

Learn the two main problems to watch out for with results based on statistical data.

Comment on how valid these claims based on statistics are:

1) Dr. Van Helsing claims that he's a more effective vampire slayer than Buffy, as he has slayed 50 vampires during his 50-year career, compared with Buffy's total (in 4 years) of 20 vampires.
2) A survey of New York motorists concluded that 85% of New Yorkers drive yellow cars.

Graphs and Charts

Make sure you know all these easy details:

1) Line Graphs or "Frequency Polygons"

A line graph or "frequency polygon" is just a set of points connected with straight lines.

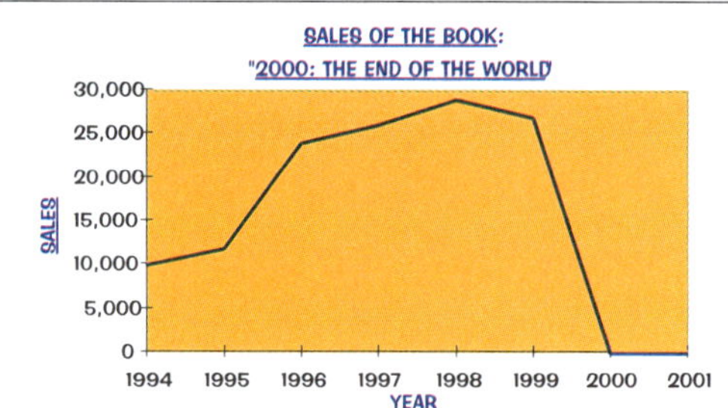

2) Bar Graphs

Just watch out for when the bars should touch or not touch:

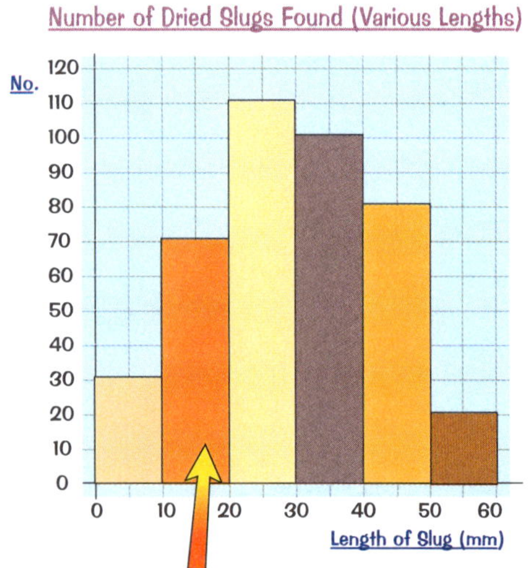

ALL the bars in this graph are for LENGTHS and you must put every possible length into one bar or the next so there can't be any spaces.

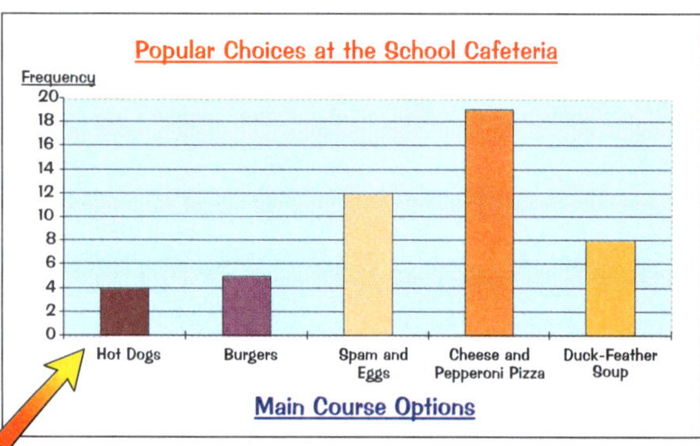

This bar graph compares totally separate items so the bars are separate.

A BAR-LINE GRAPH is just like a bar graph except you draw thin lines instead of bars.

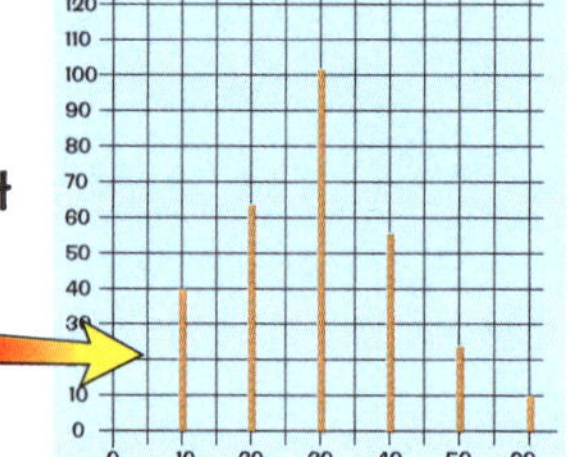

3) Stem-and-Leaf diagrams use the actual data

A stem-and-leaf diagram is a bit like a bar graph, but the data itself is used to make the diagram.

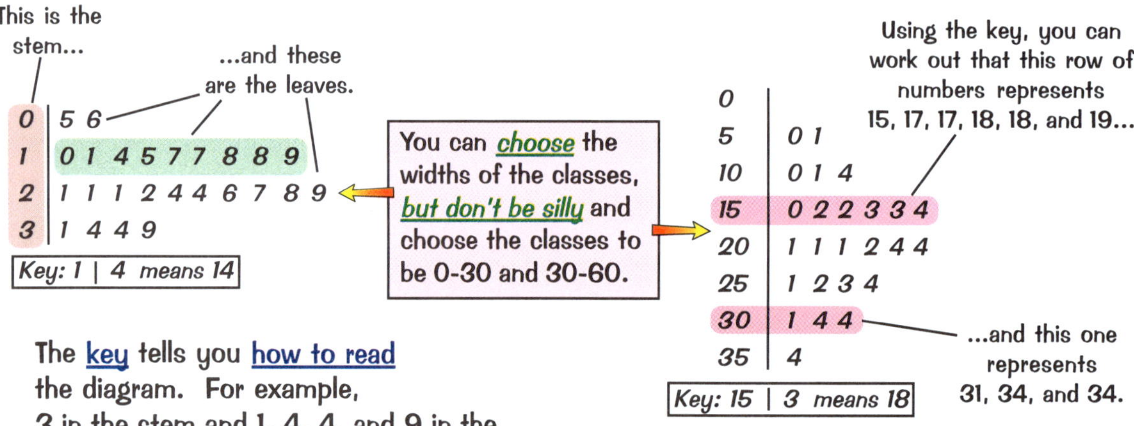

The key tells you how to read the diagram. For example, 3 in the stem and 1, 4, 4, and 9 in the leaf represent 31, 34, 34, and 39.

Graphs and Charts

4) Circle Graphs

(Also known as Pie Charts) — learn the *blindingly important* rule...

The TOTAL of Everything = 360°

Creature	Stick Insects	Hamsters	Guinea Pigs	Rabbits	Ducks	Total
Number	12	20	17	15	26	90
		×4				×4
Angle		80°				360°

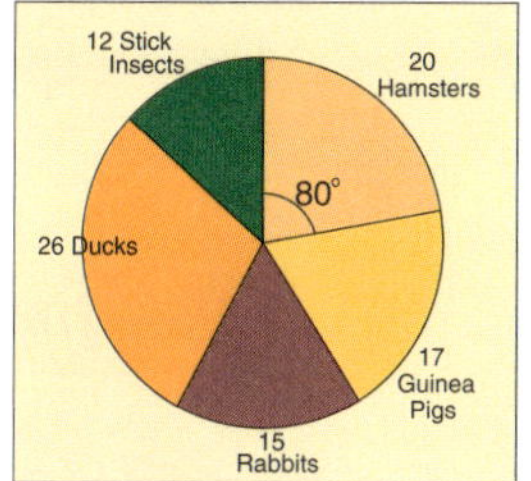

1) Add all the numbers in each sector to get the TOTAL (it's 90 for the one above).
2) Then find the MULTIPLIER (or divider) that you need to turn your total into 360°: For 90 → 360 as above, the MULTIPLIER is 4.
3) Now MULTIPLY EVERY NUMBER BY THE MULTIPLIER to get the angle for each sector, e.g. the angle for hamsters will be 20 × 4 = 80°.

5) Scatterplots — Just Draw the Points

SCATTERPLOT SHOWING THE CORRELATION BETWEEN MAX SPEED AND AVERAGE MPG FOR VARIOUS CARS
STRONG NEGATIVE CORRELATION
GOOD CORRELATION
FUEL CONSUMPTION (MPG)
MAX SPEED (MPH)

1) A SCATTERPLOT is just a bunch of points on a graph that end up in a big mess rather than in a nice line or curve.
2) There's a fancy word to say how much of a mess they're in — it's CORRELATION.
3) Good Correlation (or Strong Correlation) means the points form a nice line, and it means the two things are closely related to each other.
4) Poor Correlation (or Weak Correlation) means the points are all over the place and so there's very little relation between the two things.
5) If the points form a line sloping UPHILL (left to right), then there's POSITIVE CORRELATION, which just means that both things increase or decrease together.
6) If they form a line sloping DOWNHILL from left to right, then there's NEGATIVE CORRELATION, which just means that as one thing increases the other decreases.
7) So when you're describing a scatterplot you have to mention both things, i.e. whether it's strong/weak/moderate correlation and whether it's positive/negative.

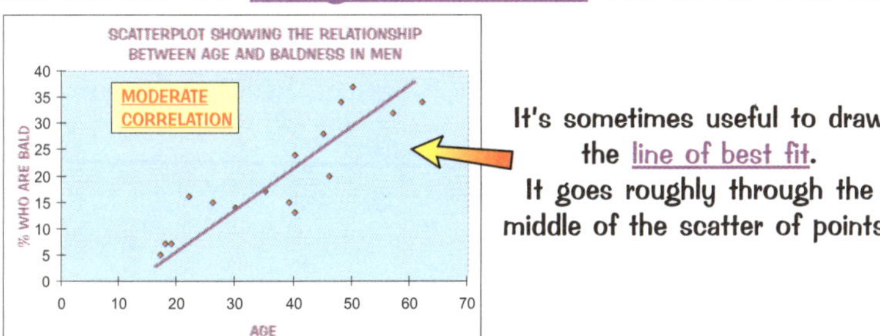

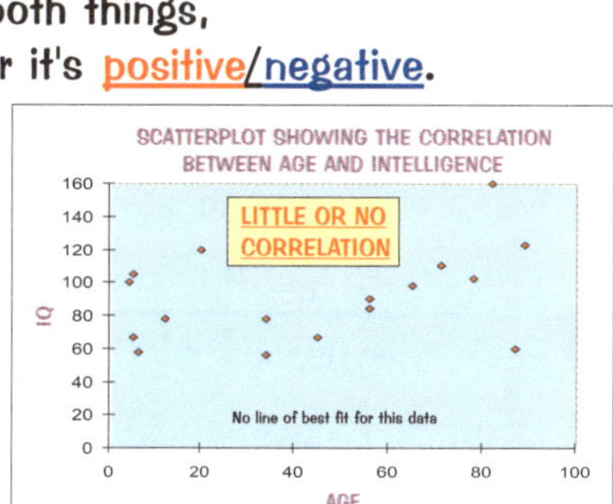

The Acid Test:

LEARN THE NAMES of the five types of GRAPHS/CHARTS.

1) Turn over the page and draw an example of each of the 5 graphs.
2) Work out the angles for all the other animals in the circle graph shown above.
3) If the points on a scatterplot are all over the place, what does it tell you about the two things that the scatterplot is showing?

Review Summary for Section Five

Section Five is the really ugly one — it's full of probability and statistics. Still, this is the last set of questions in this book, so use them wisely to test what you know. Don't forget, you've got to keep trying these *over and over again* if you want to learn this stuff properly. Just keep practicing 'til you can glide through them all like a manatee in a bikini. Or something.

Keep learning these basic facts until you know them

1) How big or small can a probability be?
2) Draw a line to represent all probabilities with words to describe it.
3) Which 3 types of numbers can be used to represent probabilities?
4) How should $P(x) = \frac{1}{2}$ be read?
5) What must the total probability add up to?
6) What is the full significance of "with replacement" or "without replacement"?
7) Explain how to find the probability of equally likely outcomes.
8) Explain how listing all possible outcomes of an event helps you find probabilities.
9) Draw a general tree diagram with all the features that tree diagrams have.
10) Give definitions for mean, median, and mode.
11) What must you do before working out the median and mode (The Golden Rule)?
12) Describe how to work out the quartiles of a set of data.
13) Explain what a box-and-whisker plot shows.
14) Which of the following can be types of untruths?
    a) lies,
    b) darned lies,
    c) statistics,
    d) all of the above.
15) Give two questions that you could ask about claims based on statistical data.
16) Describe how an average can sometimes be a more useful statistic than a total.
17) Why can results from surveys that use samples be misleading?
18) Give the names of 5 different types of charts for displaying data.
19) Draw 2 examples of each type of chart.
20) When should the bars of a bar graph touch and not touch each other?
21) Describe a method for finding the angles for a circle graph.
22) What does correlation mean? Draw graphs showing the 3 different degrees of correlation.

Answers

Section One — Numbers Mostly

P1 Multiples and Factors:

1) multiples of 7: 7, 14, 21, 28, 35, 42, 49, 56, 63, 70;
multiples of 9: 9, 18, 27, 36, 45, 54, 63, 72, 81, 90;
LCM = 63

2) factors of 36: 1, 2, 3, 4, 6, 9, 12, 18, 36;
factors of 84: 1, 2, 3, 4, 6, 7, 12, 14, 21, 28, 42, 84
GCF = 12

P2 Prime Numbers and Prime Factors:

1) 2, 3, 5, 7, 11, 13, 17, 19, 23, 29, 31, 37, 41, 43, 47

2) a) $990 = 2 \times 3 \times 3 \times 5 \times 11$ OR $2 \times 3^2 \times 5 \times 11$
b) $160 = 2 \times 2 \times 2 \times 2 \times 2 \times 5$ OR $2^5 \times 5$
c) $1260 = 2 \times 2 \times 3 \times 3 \times 5 \times 7$ OR $2^2 \times 3^2 \times 5 \times 7$

P3 Fractions:

1) a) $\frac{35}{8} = 4\frac{3}{8}$ b) $\frac{10}{33}$ c) $\frac{3}{3} = 1$ d) $\frac{4}{6} = \frac{2}{3}$

e) $\frac{3}{4} + \frac{9}{14} = \frac{3}{2\times2} + \frac{9}{2\times7} = \frac{3\times7}{2\times2\times7} + \frac{9\times2}{2\times7\times2} = \frac{21}{28} + \frac{18}{28} = \frac{39}{28} = 1\frac{11}{28}$

f) $\frac{4}{10} + \frac{4}{25} = \frac{4}{2\times5} + \frac{4}{5\times5} = \frac{4\times5}{2\times5\times5} + \frac{4\times2}{5\times5\times2} = \frac{20}{50} + \frac{8}{50} = \frac{28}{50} = \frac{14}{25}$

2) a) $3, \frac{1}{6}, \frac{7}{12}, \frac{1}{4}, \frac{1}{2}$ b) $\frac{1}{6}, \frac{1}{4}, \frac{1}{2}, \frac{7}{12}, 3$

P5 Percents:

1) Type 3, profit = \$2, 40% 2) Type 1, \$38.70
3) Type 2, \$20,500 4) Type 1, \$37.26

P6 Percents — Compound Interest:

1) \$1210 2) \$21,012.50

P7 Fractions, Decimals, and Percents:

Fraction	Decimal	Percent
$\frac{1}{5}$	0.2	20%
$\frac{7}{20}$	0.35	35%
$\frac{9}{20}$	0.45	45%
$\frac{3}{25}$	0.12	12%
$\frac{1}{8}$	0.125	12.5%
$\frac{77}{100}$	0.77	77%

P8 Squares, Square Roots, and Reciprocals:

1) a) 16 b) 196 c) 900
2) a) 7 b) 20 c) 100
3) a) 3 and 4 b) 12 and 13
4) a) $\frac{1}{12}$ b) 1 c) $\frac{7}{4}$

P9 Powers (or Exponents):

1) a) 3^8 b) 4 c) 8^{12} d) 1 e) 7^2
2) a) 5^{12} b) 6^2 c) $\frac{1}{5^2} \times \frac{1}{3^3} = \frac{1}{675}$ d) $\frac{4^2}{7^2} = \frac{16}{49}$
3) a) 2^6 b) xy^2 c) $x^{-7}y^4$

P10 Scientific Notation:

1) See P.10 2) 9.58×10^5 3) 1.8×10^{-4} 4) 4560
5) 5.0×10^{-3}, 2.5×10^{-2}, 9.1×10^6, 1.1×10^7
6) 2×10^{21}, 2,000,000,000,000,000,000,000 (21 zeros!)

P11 Negative Numbers and Multiplying Letters:

1) a) +12 (Rule 1) b) -6 (Rule1/Rule 2)
c) x (Rule 1, then Rule 2, then Rule 1) d) -3 (Rule1)
2) a) 3 b) 45 c) 11 d) 11 e) |5x| or 5|x| f) |5x| or 5|x|
3) a) 18 b) -216 c) 2 d) -27 e) -336

Section Two — Shapes and Measurement

P13 Conversion Factors:

1) 0.38 hr 2) US\$34 (= CAN\$52.70) 3) 3.2 cm

P14 Customary and Metric Units:

1) 13.3 liters 2) About 180 meters
3) 115 cm 4) 50°F

P15 Other Conversions and Scale Drawings:

1) 10.7 km/s 2) 15 inches, 1.4 miles 3) 1652.9 lb/yd^3

P16 Density and Speed:

1) See P.16 2) 16.5 g/cm^3 3) 602.25 g 4) S = D ÷ T
5) Time = $7\frac{1}{2}$ hr Dist. = 11.2 miles

P17 Density and Speed:

1) 1.875 mph 2) 4.8 g/cm^3 (= 0.0048 kg/cm^3)
3) 36 km/h

P19 Multiply for All, Then Divide:

1) 44.4 minutes 2) About 183 3) 1200 4) 11.25

P21 Perimeters and Areas:

1) Area = 154 cm^2, Circumference = 44.0 cm
2) A = 113 yd^2, C = 38 yd

P22 Perimeters and Areas:

2) Perimeter = 49.7 in, Area = 129.3 in^2

P23 Volume:

A: V = 148.5 in^3
B: V = 0.70 m^3 (Or 70,000 cm^3) — it's a cylinder.

P24 Surface Area:

1) A = 174 in^2 2) A = 4.38 m^2 (or 43,800 cm^2)

P25 Estimating:

1) Approximately 2
2) My palm's about 12 in^2 — yours is probably pretty similar to mine.
3) Approximately 500 cm^3
4) I think you'd better ask your teacher to check this one.

P26 Congruence and Enlargement:

1) 38,880 cm^3

P27 x- and y-Coordinates:

1) A(4, 5) B(6, 0) C(5, -5) D(0, -3) E(-5, -2) F(-4, 0) G(-3, 3) H(0, 5)

P28 Graph Transformations:

A→B, Reflection in the line y = x.
B→C, Translation of $\begin{pmatrix} 0 \\ 7 \end{pmatrix}$ (i.e. 7 units up).
C→D, Reflection in the line y = -x.

P29 Pythagorean Theorem:

1) BC = 8 yd, 2) 5 yd, 12 yd, 13 yd is a right triangle because $a^2 + b^2 = c^2$ *works.*

Section Three — Graphs

P33 Slope of a Straight Line:

1)

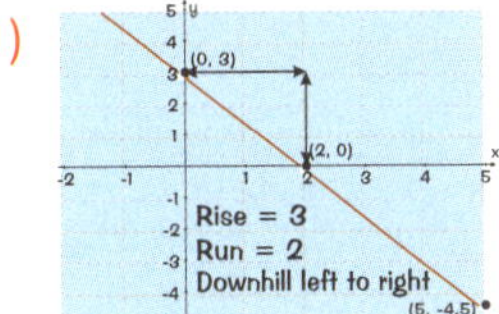

Slope = -1.5

2) Approximately 0.6 gallons per second.

Answers

P34 Graphing Straight Lines:

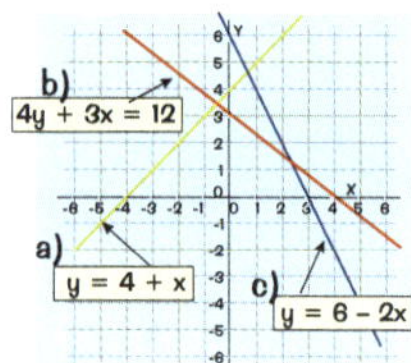

P35 Graphing Straight Lines:

1) a) $y = -2x + 3$ ($m = -2$, $b = 3$)

b) $y = -2x + \frac{1}{2}$ ($m = -2$, $b = \frac{1}{2}$)

2)

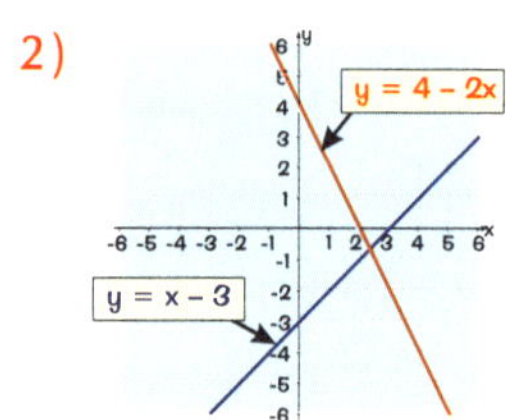

P37 Typical Questions on Straight Lines:

1) x-intercept at $x = 0.6$; y-intercept at $y = 1$

2) a) $y = -\frac{1}{2}x + 2$ b) $y = x + 3$ c) $y = -2x + 6$

3) $y = 4x + 22$ 4) $y = 4x + 59$

5) $y = 2x - 3$ and $y - 2x = 5$

P38 Curves You Should Recognize:

1)

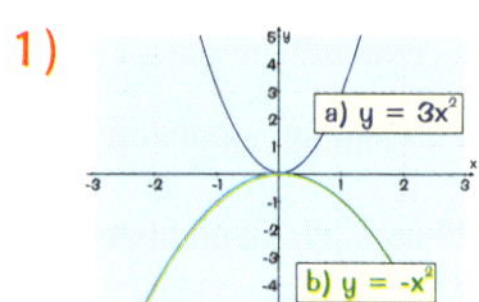

2)

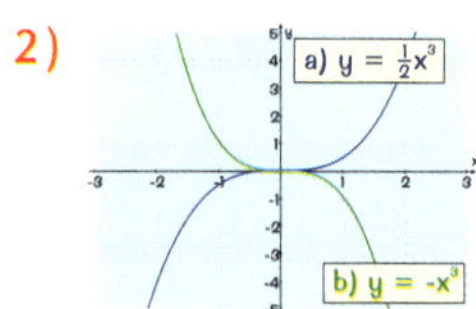

P39 Getting Answers from Your Graph:

1) a) 17 gallons b) Approximately 27 seconds

2) a) At $t = 0.7$ s and $t = 4.3$ s b) 3.6 s

P40 Travel Graphs:

1) 0.13 mph

2) Going away, stopped, going away but slower, stopped, going away but slower still, coming back very fast.

Section Four — Algebra

P42 Substituting Values into Formulas:

2) 25

P43 Solving Equations the Easy Way:

1) $x = 8$ 2) $x = 7$ 3) $x = 3$

P45 Basic Algebra:

1) a) $4x + y - 4$ b) $4y^2 - 2k + 2$ c) $\frac{2xy}{z-1}$

2) a) $6p^2q - 8pq^3$ b) $8g^2 + 16g - 10$ c) $16 - 24h + 9h^2$

3) a) $7xy^2(2xy + 3 - 5x^2y^2)$ b) $6h^2j(2j^2 + h^2jk - 6hk)$

P46 Solving Equations: a) $x = 2$ b) $x = -0.2$

P47 Rearranging Formulas:

1) $C = \frac{5}{9}(F - 32)$ 2) a) $p = -\frac{4}{3}y$ b) $p = \frac{qr}{q+r}$

P49 Word Problems:

1) \$140 2) 6 shirts 3) $l = 2w$; $2l + 2w = 100$

P50 Systems of Linear Equations: $F = 3$ $G = -1$

P51 Graphing Systems of Linear Equations:

1) a) $x = 2$, $y = 4$ b) $x = 1\frac{1}{2}$, $y = 3$

P52 Inequalities:

1) $x \geq -2$ 2) -4, -3, -2, -1, 0, 1

P53 Graphing Inequalities:

1)

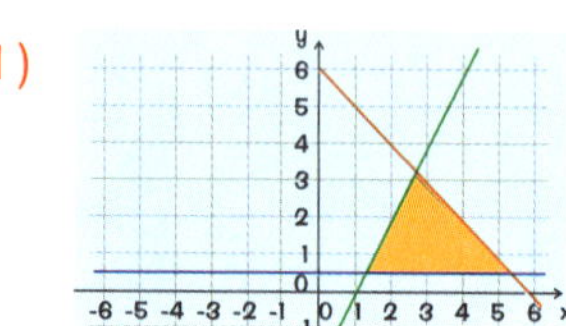

P54 Questions with Absolute Values:

1) a) $x = 4$ or $x = -4$ b) $q > 1$ or $q < -1$ c) $p > -2$ and $p < 2$

2) a) $k = 5$ or $k = 9$ b) $n > -1$ and $n < 5$ c) $m > -1$ or $m < -3$

Section Five — Probability and Statistics

P56 Probability:

1) a) $\frac{7}{20}$ b) 35% 2) $\frac{3}{4}$ 3) $\frac{1}{4}$

P57 Calculating Probabilities:

1) $\frac{2}{3}$ 2) $\frac{1}{12}$

P58 Probability — Tree Diagrams:

2) $\frac{8}{15}$

P59 Mean, Median, and Mode:

First, do this: -14, -12, -5, -5, 0, 1, 3, 6, 7, 8, 10, 14, 18, 23, 25

Mean = 5.27, Median = 6, Mode = -5

P60 Quartiles and Box-and-Whisker Plots:

Median = 67, Lower Quartile = 62, Upper Quartile = 70.5

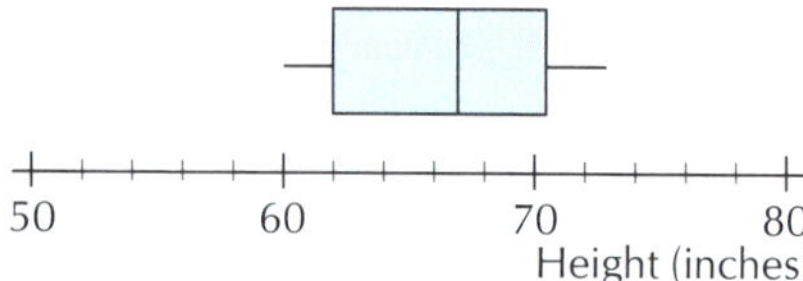

P61 Interpreting Statistics:

1) On average, Van Helsing slays 1 vampire per year, whereas Buffy slays 5 per year, so Van Helsing's claim seems unlikely to be true.

2) This is unlikely to be true. Was survey done near an airport taxi rank, or at another place where taxi cabs might be found in large numbers?

P63 Graphs and Charts:

2) Stick Insects 48°, Guinea Pigs 68°, Rabbits 60°, Ducks 104°.

3) That they are not related to each other in any way, i.e. no correlation.

Index

Index